Alina Absmeier

Spin State Switching Triggered by Light

Alina Absmeier

Spin State Switching Triggered by Light

Both Spacer Length and Parity Influence the Thermal and Light-Induced Properties of Spin Crossover Coordination Polymers

Südwestdeutscher Verlag für Hochschulschriften

Impressum/Imprint (nur für Deutschland/ only for Germany)
Bibliografische Information der Deutschen Nationalbibliothek: Die Deutsche Nationalbibliothek verzeichnet diese Publikation in der Deutschen Nationalbibliografie; detaillierte bibliografische Daten sind im Internet über http://dnb.d-nb.de abrufbar.

Alle in diesem Buch genannten Marken und Produktnamen unterliegen warenzeichen-, marken- oder patentrechtlichem Schutz bzw. sind Warenzeichen oder eingetragene Warenzeichen der jeweiligen Inhaber. Die Wiedergabe von Marken, Produktnamen, Gebrauchsnamen, Handelsnamen, Warenbezeichnungen u.s.w. in diesem Werk berechtigt auch ohne besondere Kennzeichnung nicht zu der Annahme, dass solche Namen im Sinne der Warenzeichen- und Markenschutzgesetzgebung als frei zu betrachten wären und daher von jedermann benutzt werden dürften.

Verlag: Südwestdeutscher Verlag für Hochschulschriften Aktiengesellschaft & Co. KG
Dudweiler Landstr. 99, 66123 Saarbrücken, Deutschland
Telefon +49 681 37 20 271-1, Telefax +49 681 37 20 271-0
Email: info@svh-verlag.de
Zugl.: Wien,TU,Diss.,2007

Herstellung in Deutschland:
Schaltungsdienst Lange o.H.G., Berlin
Books on Demand GmbH, Norderstedt
Reha GmbH, Saarbrücken
Amazon Distribution GmbH, Leipzig
ISBN: 978-3-8381-0226-9

Imprint (only for USA, GB)
Bibliographic information published by the Deutsche Nationalbibliothek: The Deutsche Nationalbibliothek lists this publication in the Deutsche Nationalbibliografie; detailed bibliographic data are available in the Internet at http://dnb.d-nb.de.

Any brand names and product names mentioned in this book are subject to trademark, brand or patent protection and are trademarks or registered trademarks of their respective holders. The use of brand names, product names, common names, trade names, product descriptions etc. even without a particular marking in this works is in no way to be construed to mean that such names may be regarded as unrestricted in respect of trademark and brand protection legislation and could thus be used by anyone.

Publisher: Südwestdeutscher Verlag für Hochschulschriften Aktiengesellschaft & Co. KG
Dudweiler Landstr. 99, 66123 Saarbrücken, Germany
Phone +49 681 37 20 271-1, Fax +49 681 37 20 271-0
Email: info@svh-verlag.de

Printed in the U.S.A.
Printed in the U.K. by (see last page)
ISBN: 978-3-8381-0226-9

Copyright © 2010 by the author and Südwestdeutscher Verlag für Hochschulschriften Aktiengesellschaft & Co. KG and licensors
All rights reserved. Saarbrücken 2010

Dissertation

Light Excited Spin State Switching of Spin Crossover Coordination Polymers

ausgeführt zum Zwecke der Erlangung des akademischen
Grades eines Doktors
der technischen Wissenschaften unter der Leitung von

a.o. Univ. Prof. Wolfgang Linert

am Institut für Angewandte Synthesechemie E163

eingereicht an der Technischen Universität Wien
Fakultät für Technischen Chemie
von

Dipl.-Ing. Alina Absmeier
Matr.Nr. 9825237
Lenaustr. 22
4650 Lambach

Wien, im Februar 2007

Kurzfassung

Die Eigenschaften, verbunden mit dem Einfluss von nicht koordinierenden Anionen und der Kettenlänge der verbrückenden Liganden, von Eisen(II) Spin Crossover Verbindungen wurden systematisch untersucht. Hierzu wurden Komplexe mit Tetrafluoroborat, Perchlorat bzw. Hexafluorophosphat als Gegenionen und Ditetrazolliganden mit unterschiedlicher Anzahl von C-Atomen (n = 4-10 und 12) innerhalb der Alkylenkette hergestellt und analysiert.

Die sich daraus ergebenden Veränderungen des strukturellen, optischen, magnetischen bzw. magneto-optischen Verhaltens wurden im Detail gemessen und werden in der vorliegenden Arbeit vergleichend diskutiert. Der Tetrafluoroboratkomplex mit 4 C-Atomen in der Kette, bildet ein 3 dimensionales Netzwerk und weist einen gemischten High Spin (HS) / Low Spin (LS) Zustand unterhalb von 70 K auf. Wird der Komplex innerhalb kurzer Zeit auf 4,2 K abgekühlt (gequencht), bildet sich ein metastabiler Zustand aus 80% HS und 20% LS Anteil, der bei höheren Temperaturen mit einer Halbwertszeit von 2 Tagen relaxiert. Den selben metastabilen HS Zustand kann man mit Licht geeigneter Wellenlänge (λ = 537 nm) bei 10 K anregen (LIESST-Experiment). Im Vergleich dazu zeigt der Perchloratkomplex einen 2 stufigen Übergang, mit Spin-Crossovertemperaturen bei 84 K und 134 K. Im Falle des Komplexes mit Hexafluorophosphat als Anion, wurde die Induzierung des Spinübergangs durch Magnetfelder über 100 T erstmals gemessen und präsentiert.

Komplexe mit längeren Brückenliganden (n = 5-10 und 12) kristallisieren kettenförmig wobei die Spinübergangskurven mit steigender Anzahl von C-Atomen in der Kette gradueller werden. Vergleicht man die Spinübergangstemperaturen der Perchloratkomplexe mit jenen der Tetrafluoroboratkomplexe, zeigen letztere höhere Übergangstemperaturen für 5-7 C-Atome in der Alkylenkette. Weiters teilen sich die Komplexserien abhängig von der Kettenlänge in zwei Gruppen: Solche mit geradzahliger Anzahl von C-Atomen in der Kette zeigen höhere Übergangstemperaturen als jene mit ungeradzahligen C-Atomen. Dieser „Paritätseffekt" kehrt sich in der Tetrafluoroborat Serie ab n = 8 um. Wird die Anzahl der C-Atome weiter erhöht, nähern sich die Spinübergangstemperaturen der beiden Serien 160 K. Der Paritätseffekt findet sich in der Analyse der Absorptionsspektren wieder und kann als Jahn-Teller Verzerrung interpretiert werden.

Die Ergebnisse gewähren Einblick in die sensiblen Zusammenhänge zwischen strukturellen Veränderungen und elektronischen Besetzungszuständen, welche sich in thermisch, optisch und magnetisch anregbaren Spinübergangen zeigen.

Abstract

The influence of non-coordinating anions is analysed in extension of the systematic investigations into iron(II) spin crossover coordination polymers. Ditetrazole complexes of iron(II) perchlorate and tetrafluoroborate where the two tetrazole moieties are separated by alkylene spacers are presented. The number of carbon atoms in the spacer (n) has been varied between $n = 4$-10 and 12 and the complexes have been compared. Pronounced structural, magnetic and magneto-optical changes are found in the comparison of the two series. When $n = 4$ a three dimensional network is formed that in the case of the tetrafluoroborate derivative stabilises a mixed high spin (HS)/low spin (LS) state below 70 K. If, however, the sample is quickly frozen down to 4.2 K a metastable 80% HS/20% LS state is produced, which has a $t_{1/2} = 2$ days at 44 K. In contrast the perchlorate form shows a two-step transition at 84 K and 134 K. Longer spacers ($n = 5$-10 and 12) crystallise in a chain type arrangement and decreased interactions between iron centres as n increases leads to more gradual spin transitions. When $n = 5$-7 the smaller tetrafluoroborate causes spin transitions to occur at higher temperatures than the equivalent perchlorate series. Both series exhibit a parity effect, but at $n = 8$ the parity effect reverses for tetrafluoroborate. With increasing n, however, iron-iron interactions seem to fade away and a limiting value of $T_{1/2} = 160$ K is approached by both series. This observation of the parity effect can also be proved when analysing the absorption spectra of the series. The magnetic field induced high-spin low-spin transition of [Fe(4*ditz*)$_3$](PF$_6$)$_2$ in transient megagauss fields using optical reflection measurements is described.

Danksagung

An erster Stelle möchte ich mich ganz besonders bei Prof. Dr. Wolfgang Linert (TU Wien) für die Themenstellung und für die Möglichkeit an diesem Projekt mitarbeiten zu können bedanken.

Großen Dank möchte ich ebenfalls Dr. Guy N. L. Jameson (University of Otago, NZ) aussprechen, ohne dessen Hilfe und Engagement die Auswertung der Daten in diesem Ausmaß nicht möglich gewesen wäre.

Gleiches gilt für Dr. Franz Werner (TU Wien) der sich jederzeit unserer Proben angenommen hat und die "Nichtwisssenden" mit sehr viel Geduld in die Materie "Pulver" eingeführt hat.

Ein herzliches Dankeschön gebührt ebenfalls Dr. Peter Weinberger (TU Wien), für die organisatorische Betreuung und wissenschaftliche Beratung.

Aufs herzlichste möchte ich mich auch bei Prof. Dr. Jean-Francois Létard (University of Bordeaux) und Dr. Chiara Carbonera (University of Zaragoza), für die exzellente Betreuung und erhaltenen Ergebnisse bedanken.

Ebenso ist Prof. Dr. Michael von Ortenberg (Humboldt Universität zu Berlin) und seiner Assistentin Bettina Richter für die hervorragende Zusammenarbeit zu danken.

Für die Durchführung der Mössbauer wie auch magnetischen Messungen, möchte ich Prof. Dr. Michael Reissner und Prof. Dr. Günter Wiesinger herzlich danken.

Prof. Dr. Kurt Mereiter (TU Wien) sowie Prof. Sally Brooker (University of Otago, NZ) danke ich für die Messungen der Kristallstrukturen.

Ein Dankeschön möchte ich Hr. K. Kato und Hr. M. Takata (SPring-8, Sayo-gun, Japan) für die Durchführung der Synchrotron Messungen aussprechen.

Bei Frau Poppenberger (TU Wien) sowie Hr. Hasegawa (Aoyama-Gakuin University, Japan) möchte ich mich für die Durchführungen der SEM Messungen bedanken.

Für allerlei hilfreiche Gespräche und Anregungen während der Durchführung der Arbeiten, möchte ich allen voran Dipl.-Ing. Matthias Bartel danken, der mir stets mit hervorragenden wie auch erheiternden Ratschlägen zur Seite stand. Ebenfalls danke ich an dieser Stelle den verbleibenden Kollegen der Forschungsgruppe für Koordinationschemie und Bioanorganische Chemie, Dipl.-Ing. W. Freinbichler und Nader Hassan, für den hilfreichen Informationsaustausch und das hervorragende Arbeitsklima.

In diesem Zusammenhang möchte ich den *"Kaffee - Kollegen"* für ihre regelmäßigen Besuche und die damit verbundenen Pausen und erörterten Erkenntnisse danken.

Nicht zuletzt möchte ich mich besonders bei meinen Eltern, wie auch allen Verwandten, Bekannten und Freunden, für ihre ständige Unterstützung während meines Doktorates und ihrem offenen Interesse an meiner Arbeit herzlichst bedanken.

Contents

1.
   Introduction   2
   1.1. First Investigations 2
   1.2. Theoretical Background 3
   1.3. Possibilities to Induce Spin Transition 5
      1.3.1. Spin Transition Induced by Temperature 5
      1.3.2. Spin Transition Induced by Pressure 7
      1.3.3. Spin Transition Induced by Light 8
      1.3.4. Spin Transition induced by a High Magnetic Field 10
   1.4. Heterocyclic Ligands 12

2.
   Experimental Part   16
   2.1. Chemicals and standard physical characterization 16
   2.2. Synthesis and Crystal Structure of the Ligands 17
   2.3. Synthesis of the Complexes 22
   2.4. Laboratory X-ray Powder Diffraction 27
   2.5. Synchrotron Powder Diffraction 27
   2.6. Scanning Electron Microscope (SEM) 27
   2.7. Optical Measurements 28
      2.7.1. UV/VIS-NIR Measurements 28
      2.7.2. Reflectivity Measurements 28

Contents

- 2.8. Magnetic Measurements 29
    - 2.8.1. Magnetic measurements 29
    - 2.8.2. Magneto-optical Measurements 30
- 2.9. Mössbauer Spectroscopy 31
- 2.10. High magnetic field 32
    - 2.10.1. Generation of Magnetic Fields 32
    - 2.10.2. Detection of the Magnetic Field 35
    - 2.10.3. Sample Holder and Cooling System 36

3. Results   38

- 3.1. Powder Diffraction 38
    - 3.1.1. Synchrotron Powder Diffraction 44
- 3.2. UV/VIS-NIR Spectroscopy 46
    - 3.2.1. Analysis of the Absorption Spectra of the BF_4^- and ClO_4^- Series 47
    - 3.2.2. Influence of the Anions 54
    - 3.2.3. Solvent Effect 56
- 3.3. Reflectivity Measurements 57
- 3.4. Far-FTIR properties of the ClO_4^- series 62
- 3.5. Magnetic Measurements 63
    - 3.5.1. [Fe(n*ditz*)$_3$](ClO$_4$)$_2$ serie 63
    - 3.5.2. [Fe(n*ditz*)$_3$](BF$_4$)$_2$ series 66
- 3.6. Photo-Magnetic Measurements 68
    - 3.6.1. LIESST Properties of the ClO_4^- Complexes 68
    - 3.6.2. LIESST Properties of the BF_4^- Series 71
- 3.7. Spin Crossover in Megagauss Fields 75
    - 3.7.1. Constant Field Measurements 77
    - 3.7.2. Variation of the Magnetic Field 78

4.
    Discussion   81
    4.1. Powder Diffraction 81
    4.2. UV/VIS-NIR Spectroscopy 84
    4.3. Magnetic and Photo-Magnetic Measurements 89
    4.4. High Magnetic Field Measurements 98
        4.4.1. Simulation 98

5.
    Conclusion   102

A.
    Reflectivity   107

B.
    Publications   112

Bibliography

List of Figures

List of Tables

Introduction

1. Introduction

1.1. First Investigations

In the 1930s Cambi *et al.* observed for the first time the anomalous magnetic properties of some N,N'-substitueted tris(dithiocarbamate)iron(III) complexes [1]. Concurrently, Pauling and co-workers discussed similar anomalous magnetic behaviour of ferrihaemoprotein hydroxides [2]. Around twenty years elapsed between these investigations and the time when the foundations of the ligand field theory were solidly established. At this time, Orgel suggested the possibility for a spin state equilibrium for these anomalous properties [3]. Subsequently, Griffith and co-workers observed the expected thermal equilibrium between spin states for several haemoproteins [4] and Martins, White and co-workers published the first theoretical interpretation of the magnetic behaviour in the tris(dithiocarbamate)iron(II) complex [5]. In 1964 Baker and Bobonich reported an unusual cooperative behaviour observed for the complexes [Fe (*phen*)$_2$ (NCX)$_2$] (X=S,Se) and [Fe (*bipy*)$_2$ (NCS)$_2$], which represented the first Fe(II) spin-crossover (SC) systems (whereas the expression of "spin-crossover" was established by Ewald *et al.* in the same year [5]). However, they did not associate these observations with a spin transition between S=0 and 2 spin states for iron(II) [6]. Three years later König and Madeja established definitively the nature of the spin transition for these iron(II) derivatives from detailed magnetic and Mössbauer spectroscopic studies [7]. Soon after this fundamental results the number of SC compounds with different ions raised rapidly. So that the research work in this field is nowadays distributed on the following transition metals: Fe(III) [1, 8], Cr(II) [9], Mn(II) [10], Co(II) [11, 12], Mn(III) [13], Co(III) [14] and of course Fe(II) about which this work is focused.

1.2. Overview of the Theoretical Background

In octahedral symmetry (O_h) iron(II) complexes can adopt two different electronic ground states depending to a first approximation on the magnitude of the Δ energy gap between the e_g and t_{2g} metal d-orbitals relative to the mean spin pairing energy P. More precisely, for $\Delta \gg P$, the ground state arises from the configuration where the d electrons occupy first the t_{2g} orbitals of lowest energy. The ground state is then LS. For $\Delta \ll P$, on the other hand, Hund's rule is obeyed. The HS ground state has the same spin multiplicity as the free metal ion. When the conditions $\Delta \ll$ or $\gg P$ are no longer fulfilled, a LS $\leftrightarrow$ HS transition may occur. The condition for a spin transition to occur is often defined by $|\Delta - P| \approx kT$. Such a formulation is often confusing. As a matter of fact, Δ depends on the electronic state: for a SC compound Δ(HS) and Δ(LS) are clearly not the same. Similarly, the mean spin pairing energy P is defined in rather ambiguous way. The occurrence of spin transition is related to LS and HS potential energy curves (see Figure 1.1). A spin-transition may be induced in a controlled, detectable and reversible manner, by the action of temperature (T), pressure (p), magnetic field (h) or light irradiation (hν) (see Figure 1.1).

The SC phenomenon can be considered to be an intra-ionic electron transfer, where the electrons move between the e_g and the t_{2g} orbitals. Given that the e_g subset has an antibonding character its population/depopulation takes place concomitantly with an increase/decrease in the metal-to-ligand bond distances. An opposite change in the population takes place in the t_{2g} orbitals, which affects the electron back-donation between the metal ion and the vacant π^* orbitals of the ligands. Both σ and π^* factors contribute to the change of the metal-ligand bond length. The typical mean metal-to-ligand bond length change, ΔR, is around 0.2 Å for Fe(II). Hence, a remarkable change of the molecular size and shape takes place during the spin conversion (see Figure 1.1). It is worth noting, that in addition to the metal-ligand bond length changes, remarkable variations of the bond angles are also observed. Consequently the

1.2. THEORETICAL BACKGROUND

SC molecule experiences a drastic change of Δ_O upon spin conversion, which may be estimated from $\Delta_{LS}/\Delta_{HS} \approx (\Delta R_{HS}/\Delta R_{LS})^n$ with $n = 5\text{-}6$. For instance, $\Delta_{LS} \approx 1.75\ \Delta_{HS}$ for Fe(II). This estimation neglects the angular dependence of Δ_O and considers that ΔR is the most significant structural parameter as confirmed experimentally [15]. With these theoretical background in mind, many efforts have been made to design "molecular materials" possessing a set of properties tunable by external constraints with the target to reproduce at the molecular scale traditional electronic functions, like memories, modulators, rectifiers, transistors, switches and wires [16].

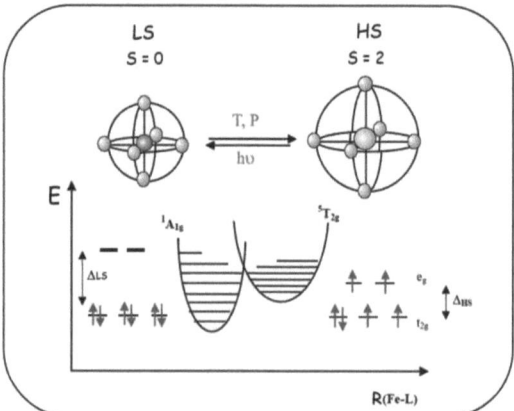

Figure 1.1.: Presentation of the adiabatic potentials for the high-spin and the low-spin state along with the most important reaction coordinate for SC in Fe(II) complexes.

1.3. Possibilities to Induce Spin Transition

1.3.1. Spin Transition Induced by Temperature

Temperature is historically the most commonly used method to induce the spin transition [1, 5, 8, 9, 10, 11, 12, 13, 14, 17, 18, 19]. It is this method which all the others are established on.

As spin conversions have been observed in the solid state as well as in solution, Ewald *et al.* [5] discovered for the latter case, that the process is essentially molecular, and the spin transition is very gradual, obeying a Boltzmann distribution law between the two spin states. On the macroscopic scale, the situation is more complicated and the spin transition has to be described thermodynamically. The change of the spin state is given by a physical equilibrium between the LS and the HS state which can be changed by the variation of the Gibbs free energy ΔG or the entropy factor ΔS and the enthalpy factor ΔH (see equation 1.1).

$$\Delta G = G_{HS} - G_{LS} = \Delta H - T\Delta S \tag{1.1}$$

For example, the variation of ΔH is directly connected with the electronic contribution, ΔH_{el} [20] which is positive for the LS $\rightarrow$ HS transition. The entropy, which is the sum of electronic ΔS_{el} and vibrational ΔS_{vib} contributions, is also positive for the LS $\rightarrow$ HS transition. If ΔH and ΔS have the same sign, there exists an equilibrium temperature $T_{1/2}$. This critical temperature, at which the same amount of LS and HS molecules exists, is defined by $\Delta G = 0$, hence

$$T_{1/2} = \Delta H/\Delta S \tag{1.2}$$

Below $T_{1/2}$ ΔH is greater than $T\Delta S$ ($\Delta G \geq 0$) and therefore the LS is more stable. On the opposite, above $T_{1/2}$ ΔH is smaller than $T\Delta S$ ($\Delta G \leq 0$) and the HS state is more stable. An increase in temperature therefore stabilizes the HS state because of a gain of entropy during the LS $\rightarrow$ HS transition. The thermal spin

1.3. POSSIBILITIES TO INDUCE SPIN TRANSITION

transition is consequently a molecular process dominated by entropy [21, 22]. Until 1964, the obtained spin transition curves of Fe(II) compounds in the liquid phase were all gradual. It is also possible to obtain such curves in the solid state, but because of possible cooperative interactions in the solid phase, the transition curve can take many shapes. In 1964 the first compound which showed an abrupt temperature dependent spin transition was [Fe(*phen*)$_2$(NCS)$_2$] [19], in 1976 the [Fe (4,7 − (CH$_3$)$_2$ − *phen*)$_2$ (NCS)$_2$] was the first compound with an hysteresis, measured by König and Ritter [23]. Today there are a lot of compounds known which show differences between 2 and 90 K when they are cooled down and afterwards heated again [24], whereas for example an hysteresis of 50 K can be attributed to a loss of solvent (often water) during the first temperature cycle [25, 26]. In 1982 the first two step spin transition was observed with the compound [Fe (2 − *pic*$_3$)] Cl$_2$ · EtOH [27]. The trends of these curves are shown schematically in Figure 1.2.

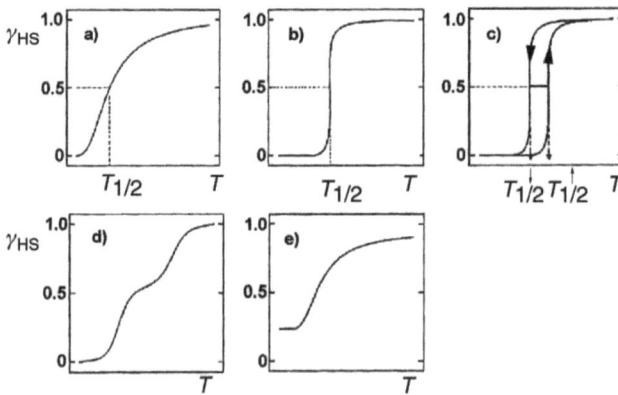

Figure 1.2.: Principal types of spin-transition curves represented in the form of high-spin molar fraction, γ_{HS}, vs. temperature T: (a) gradual, (b) abrupt, (c) abrupt with hysteresis, (d) two-step and (e) incomplete.

1.3. POSSIBILITIES TO INDUCE SPIN TRANSITION

1.3.2. Spin Transition Induced by Pressure

The effects of applied pressure on the SC have already been studied by Ewald *et al.* [5]. As explained under 1.2 the LS state has a smaller molecular volume than the HS state and is favoured as pressure increases. This means that the spin transition occurs at higher temperatures compared to the spin transition temperature at normal pressure (see Figure 1.3).

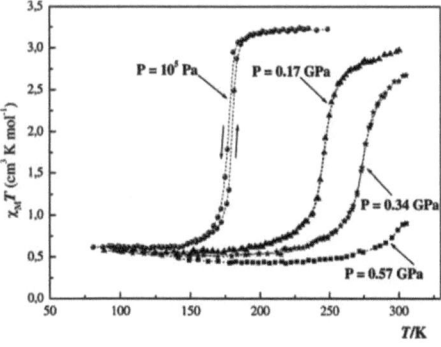

Figure 1.3.: $\chi_M T$ vs. T curves at different pressures for [Fe(*phen*)$_2$(NCS)$_2$] polymorph II [33].

Systematic and detailed studies of the concerted action of temperature and pressure variation of SC compounds have only recently become possible with the development of special hydrostatic pressure cells in connection with magnetic susceptibility, optical and Mössbauer measurements, EXAFS and vibrational spectroscopy [28, 29]. After this technical progress significant results of pressure effect studies on solid mononuclear, dinuclear and polymeric 1D, 2D and 3D SC compounds of iron(II) have been obtained [30, 31]. These results led to important information on thermodynamical properties and on the driving force of the thermal spin transition in SC compounds. Analysing the evolution of the spin transition curve obtained under pressure allows to extract, along with the changes of enthalpy and entropy, the change of the volume of the

1.3. POSSIBILITIES TO INDUCE SPIN TRANSITION

unit cell upon spin conversion as well as the behaviour of the interaction constant. Application of pressure is also a powerful tool to investigate the role of the structure and the interplay between spin transition and structural phase transition [29, 32].

1.3.3. Spin Transition Induced by Light

The first evidence of **L**ight-**I**nduced **E**xcited **S**pin **S**tate **T**rapping - LIESST effect - in an iron(II) SC material was reported by McGarvey and Lawthers [34] in solution and then by Decurtins *et al.* [35] in the solid state. The latter authors demonstrated the possibility of converting a LS state into a metastable HS state at low temperatures ($\leq$ 50 K) by using green light irradiation. Later Hauser [36] showed that red laser light switches the system back to the LS state (reverse - LIESST effect). The mechanism for these photo-switching processes, which turned out to be a common feature of most Fe(II) SC systems, is shown in Figure 1.4.

Green light (514 nm) is used for the spin allowed excitation $^1A_1 \rightarrow {}^1T_1$ with $^1T_1$ lifetimes typically of nanoseconds. A fast relaxation cascading over two successive intersystem crossing steps, $^1T_1 \rightarrow {}^3T_1 \rightarrow {}^5T_2$, populates the metastable $^5T_2$ state. Radiative relaxation $^5T_2 \rightarrow {}^1A_1$ is forbidden, and decay by thermal tunneling to the ground state $^1A_1$ is slow at low temperatures. Reverse-LIESST is achieved by application of red light ($\approx$ 820 nm) whereby the $^5T_2$ is excited to the ^5E state with the two subsequent intersystem crossing processes, $^5E \rightarrow {}^3T_1 \rightarrow {}^1A_1$, leading back to the LS ground state. As demonstrated later by Hauser, photoswitching is also possible via $^1A_1 \rightarrow {}^3T_1 \rightarrow {}^5T_2$ transitions using red light of 980 nm. The lifetime of the photoinduced HS state is usually long at very low temperature, *e.g.* weeks at 20K for [Fe(*ptz*)$_6$](BF$_4$)$_2$ [36], but above 50 K the relaxation process becomes thermally activated and in few seconds the stored light induced information has vanished. In 1991 Hauser introduced the first guideline allowing some expectations of the lifetime of the photoinduced HS state [38]. By carefully investigating the dynamics of the LIESST effect in different diluted SC compounds he noticed a strong correlation between

1.3. POSSIBILITIES TO INDUCE SPIN TRANSITION

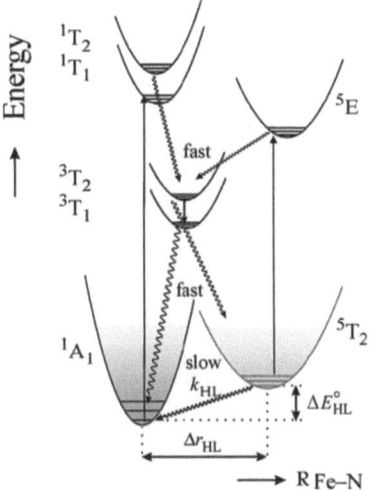

Figure 1.4.: Schematic illustration of LIESST and reverse- LIESST of a d^6 complex in the SC range. Spin allowed d-d transitions are denoted by arrows and the radiationless relaxation processes by waved lines [37].

the lifetime of the LIESST state extrapolated to $T \to 0$ (e.g. in the tunneling region [39]), expressed as $lnk_{HL}(T \to 0)$, and the thermal SC temperature, $T_{1/2}$. In other words, Hauser demonstrates that the lifetime of the metastable LIESST state is inversely proportional to $T_{1/2}$. This relation is today known as the *inverse - energy - gap* law. Based on this result Létard et al. measured more than sixty SC compounds and summarised the results in the so called *T(LIESST)* database, which clarifies the *T(LIESST)* / $T_{1/2}$ relation [40].

The relaxation of metastable LIESST states applies also to the relaxation of metastable spin states generated by "nuclear decay-induced excited spin state trapping" (NIESST) [28]. It can therefore be concluded that both phenomena, LIESST after optical excitation and NIESST after nuclear decay, follow the same relaxation mechanism.

1.3. POSSIBILITIES TO INDUCE SPIN TRANSITION

The effect of light on SC systems is being pursued very actively and a number of new phenomena have recently been reported. These include: **L**ight-**I**nduced **T**hermal **H**ysteresis (LITH), the effect of generating an hysteresis in the spin transition curve under constant irradiation [41]. **L**ight-**I**nduced **P**erturbation of a **T**hermal **H**ysteresis (LIPTH), whereby the hysteresis associated with a thermal spin transition is shifted to either lower or higher temperatures under light irradiation with light of different wavelengths [42]. **L**igand-**D**riven-**L**ight **I**nduced **S**pin **C**hange (LD-LISC) is another remarkable light effect, whereby irradiation affords a *cis-trans* isomerisation of the coordinated ligand, *e.g.* of stilbenoid type, with subsequent spin transition at the Fe(II) centre as a consequence of a change in ligand field strength [43].

1.3.4. Spin Transition induced by a High Magnetic Field

Keeping in mind that (i) a magnetic field stabilises the highest spin-state of the molecule through the Zeeman effect, (ii) the SC phenomenon is a dynamic process and (iii) it is possible to generate high pulsed magnetic fields the possibility to induce the SC by applying a magnetic field was first demonstrated by Sasaki and Kambara, who proposed a model based on the ligand field theory, predicting the effect of large static magnetic fields (20-200 T) in ferrous and ferric compounds [44]. Shortly after, Qi *et al.* [45] experimentally quantified the effect for a lower static field (5.5 T) on the iron(II) SC compound [Fe(*phen*)$_2$(NCS)$_2$], which revealed a shift of the thermal spin transition by -0.12 K. In 2000 Bousseksou *et al.* [46] reported the first study of the effect of a high and pulsed magnetic field (32 T) on the spin state of the above mentioned compound. A partial triggering of the SC was demonstrated with at least a 15% HS fraction ceated in an irreversible process. A typical result of these experiments is given in Figure 1.5.

In the future prospects of designing molecular switching devices based on SC compounds, the use of pulsed magnetic fields for the investigation of the switching dynamics represents an appealing perspective. Until now, exper-

1.3. POSSIBILITIES TO INDUCE SPIN TRANSITION

iments on SC complexes consisted of applying 32 T pulsed magnetic fields of about 1 s [46] are known. At present, microsecond magnetic fields well beyond this value are available using high-field facilities, which opens new perspectives for fast triggering of spin transition. Fields above 100 T and with durations of few microseconds can be reached in the Humboldt Magnetic Field Center in Berlin and Tokyo University, the absolute record being 2800 T was obtained at the Russian Federal Nuclear Center (VNIEF).

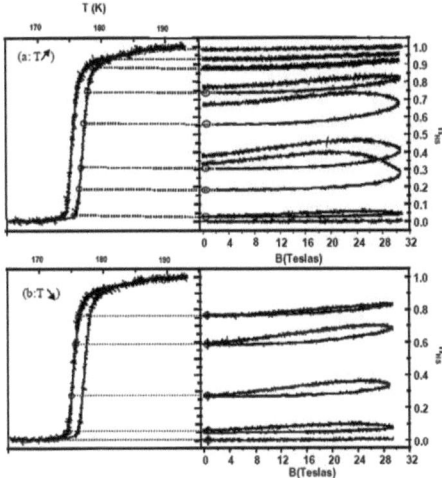

Figure 1.5.: [Fe(*phen*)$_2$(NCS)$_2$], the complete set of pulsed field experiments in the ascending (a) and descending (b) branches of the thermal hysteresis loop [46].

1.4. Spin Crossover Complexes with Heterocyclic Ligands

Iron(II) SC compounds have been produced with many different 'N' coordinating heterocyclic ligands, including triazoles [47], tetrazoles [48, 49, 50, 51, 52] and imidazoles [53]. In particular, tetrazole containing ligands have been successfully employed as part of a family [54] of terminal and bridging ligands to form 1D chain structures [55, 56] and 3D networks [52]. The latter, [μ -tris-(1,4-bis(tetrazol-1-yl)butane-N4,N4´)iron(II)]-bis(hexafluorophosphate) ([Fe(4$ditz$)$_3$](PF$_6$)$_2$) synthesised from MeOH, proved to be particularly interesting. Each iron(II) is octahedrally coordinated by symmetry equivalent tetrazole rings with the butylene spacer outstretched, assuming a zigzag configuration spanning iron(II) centres (see Figure 1.6). The ditetrazole ligands link the Fe(II) atoms into a 3D network and three such 3D networks interpenetrate each other. This structure results in cooperativity between iron(II) centres as borne out by SQUID measurements. The compound has a two-step spin transition at 168 K (with an hysteresis of 5 K) and 173 K [52]. While the shorter ethylene bridged ligand, 1,2-bis(tetrazol-1-yl)ethane, produces a 1D polymeric chain structure [56] with a more gradual spin transition curve than the butylene analogue.

With these results in mind the recent research efforts are focused onto ditetrazole ligands with various length alkylene spacers between the coordinating tetrazole groups (see Figure 1.6).

Therefore in the first part of this work the empahsis is placed on two series of complexes, namely the [μ-tris-(1,n-bis(tetrazol-1-yl)butane-N4,N4´)iron(II)]-bis(hexafluoroborate) [Fe(n$ditz$)$_3$](BF$_4$)$_2$ and the [μ-tris-(1,n-bis(tetrazol-1-yl)butane-N4,N4´)iron(II)]-bis(perchlorate) [Fe(n$ditz$)$_3$](ClO$_4$)$_2$ with n = 4 - 10 and 12. The idea was to study systematically the influence of the number of carbons (n) in the spacer as well as the influence of the non-coordinating anion on the magnetic, photomagnetic and structural properties of the compounds.

1.4. HETEROCYCLIC LIGANDS

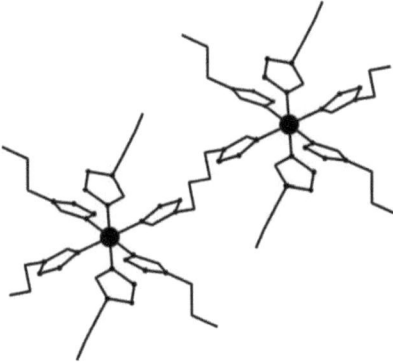

Figure 1.6.: The basic structural units of [Fe(4*ditz*)$_3$](PF$_6$)$_2$ showing how the ligands span the iron(II) centres to produce a three dimensional polymer.

To make a reasonable comparison two conditions had to be fullfield:
(i) non-coordinating anions, tertafluoroborate (BF$_4^-$) and perchlorate (ClO$_4^-$), of similar size (estimated ionic radii are 2.68 Å for BF$_4^-$ and 2.78 Å for ClO$_4^-$, geometry and chemical hardness were used and (ii) the complexes were all prepared identically. All compounds of the two series have been characterized by temperature dependent UV/Vis-NIR spectroscopy, reflectivity experiments, SQUID, LIESST and X-ray powder Diffraction (XRPD).

During the work it became apparent that the [Fe(4*ditz*)$_3$](X)$_2$ with X = BF$_4^-$, ClO$_4^-$, PF$_6^-$ (2.95 Å), ReO$_4^-$ (3.08 Å) and SbF$_6^-$ (3.17 Å) complexes are structurally different and show different magnetic behaviour in comparison to the complexes with longer chains. Therefore the change in the ligand field parameters, depending on the size of the anions[1], were studied in detail using temperature dependent UV/VIS-NIR spectroscopy.

Responding to a growing interest in compounds which exhibit the LIESST effect additionally to the above mentioned series the [Fe(4*ditz*)$_3$](PF$_6$)$_2$, [Fe(n*ditz*)$_3$](PF$_6$)$_2$ with n = 4, 7-9 and [Fe(n*ditz*)$_3$](SbF$_6$)$_2$ with n = 7-9 com-

---
[1] The size of the anions is dependent upon the chemical surrounding. To be able to compare the size of the anions with the magnetic behaviour and to be consistent, the size was estimated via the effective ionic radii table presented by Shannon [57]

1.4. HETEROCYCLIC LIGANDS

plexes have been checked for their LIESST properties.

Another interest of this work is the measurement of SC properties under different pulsed high magnetic fields. One expects some important information concerning the dependence between the spin relaxation time as well as the reaction time of the SC at different magnetic fields. Therefore the behaviour of the [Fe(4*ditz*)$_3$](PF$_6$)$_2$.EtOH complex when applying fields of 140 T at different temperatures of the SC curve.

Experimental

2. Experimental Part

2.1. Chemicals and standard physical characterization

L-ascorbic acid, 1,5-diamino-pentane ($\leq$ 98 %), 1,7-diamino-heptane (99 %), 1,9-diamino-nonane (99 %), glacial acetic acid (99 %), iron(II)perchlorate-hexahydrate, iron(II)tetrafluoroborate-hexahydrate, sodium azide, sodium hydroxide (97 %) and triethyl orthoformate were obtained from Aldrich. All other chemicals were standard reagent grade and used as supplied.

Elemental analysis (C, H and N) were performed by the Mikroanalytisches Laboratorium, Faculty of Chemistry, Vienna University, Währingerstrasse 42, A-1090 Vienna, Austria and in the laboratories of Dr. Roman Boca, Slovenská Vyská Scola Technicá v Bratislave, Chemitechothenologická Fakulta, Radlinského 9, SK-81237 Bratislava, Slovakia.

Mid-range FTIR spectra of the compounds were recorded as KBr-pellets within the range of 4400 - 450 cm^{-1} using a Perkin-Elmer 16PC FTIR spectrometer. Pellets were obtained by pressing the powdered mixture of the samples in KBr in vacuo using a hydraulic press applying a pressure of 10.000 $kgcm^{-2}$ for 5 minutes.

Far-range FTIR spectra were recorded within the range 600 - 250 cm^{-1} on a Perkin-Elmer System 2000 Far-FTIR spectrometer. The complexes were diluted with polyethylene and pressed with a pressure of 10.000 $kgcm^{-2}$ transiently. Variable temperature Far-IR spectra in the temperature range

100 - 298 K were recorded using a Graseby-Specac thermostatable sample holder with polyethylene windows, attached to a Graseby-Specac automatic temperature controller.

^{1}H-NMR and ^{13}C-NMR in deuterated [d^6]DMSO were measured using Bruker DPX 200 MHz and Bruker 250 FS FT-NMR spectrometers. Proton NMR chemical shifts were reported in ppm calibrated to the respective solvent.

2.2. Synthesis and Crystal Structure of the Ligands

WARNING: Tetrazole and perchlorate compounds should be handled with care; they may detonate upon heating or shock!!!

The general synthetic pathway and the synthesis of the even-numbered ligands 4*ditz*, 6*ditz*, 8*ditz*, 10*ditz* and 12*ditz* has been reported in the literature [54, 58, 59]. Modified procedures were used to produce three odd-numbered ligands 1,5 -bis (tetrazol - 1 - yl) pentane [5*ditz*], 1,7 - bis (tetrazol - 1 - yl) heptane [7*ditz*] and 1,9 - bis (tetrazol - 1 -yl) nonane [9*ditz*] as given below (Note that the 1,11 - bis (tetrazol - 1 - yl) undecan [11*ditz*] has not yet been synthesized):

80 mmol of the respective diamines, 160 mmol sodium azide and 160 mmol triethyl ortho formate were stirred in a 500 ml three-necked round bottom flask. 250 ml 99.5% acetic acid was added and heated to 90 - 95°C for four hours. After 4 and 16 hours reaction time another 160 mmol triethyl orthoformate and 160 mmol sodium azide were added and stirred for an additional 24 hours at 95°C. After cooling the reaction mixture was poured into a beaker and a saturated sodium hydrogencarbonate solution was added under vigorous stirring to neutralise the acetic acid, followed by solid sodium hydrogencarbonate to precipitate the product. The suspension was cooled to 4°C for 3 hours and the precipitate was filtered off and recrystallised from ethanol. The colourless needle-shaped crystals were dried over P_2O_5. Analytical data are presented in

2.2. SYNTHESIS AND CRYSTAL STRUCTURE OF THE LIGANDS

Table 2.1.

The single crystals used for X-ray diffraction were obtained via solvent evaporation from ethanol (5*ditz*), pyridine (7*ditz*) and DMF (9*ditz*).

Crystals of 5*ditz*, 7*ditz* and 9*ditz* were all elongated lath-like and soft. Selected crystals were mounted on a Bruker SMART diffractometer (graphite monochromated Mo-Kα radiation from a sealed X-ray tube, λ = 0.71073 Å platform 3-circle goniometer, CCD area detector) and intensity data were collected at room temperature. After raw data extraction with program SAINT, absorption and related effects were corrected with program SADABS (multi-scan method) and data were processed with program XPREP [60]. The structures were then solved with direct methods using SHELXS97 followed by structure refinements on F^2 with program SHELXL97 [61]. Non-hydrogen atoms were refined anisotropically. Hydrogen atoms were inserted in calculated positions and refined with the riding model. Crystallographic data are given in Table 2.2 and ORTEP plots are given in Figure 2.1 with the corresponding even ligands. CCDC 268788 - 268790 contains the supplementary crystallographic data. These data can be obtained free of charge from The Cambridge Crystallographic Data Centre via *www.ccdc.cam.ac.uk/data_request/cif*.

2.2. SYNTHESIS AND CRYSTAL STRUCTURE OF THE LIGANDS

Table 2.1.: Yield, NMR and IR data of 5*ditz*, 7*ditz* and 9*ditz*.

	5*ditz* $C_7H_{12}N_8$	7*ditz* $C_9H_{16}N_8$	9*ditz* $C_{11}H_{20}N_8$
Yield [g, %]	3.22, 10	1.2, 6.4	3.5, 22.3
m.p. [°C]	125 – 127	85 – 86	92 – 93
^{1}H-NMR (250 MHz, d^6 DMSO) [ppm]	9.39 (s, 2H) 4.44 (t, 4H, J=7.03 Hz) 1.86 (quin, 4H, J=7.19 Hz) 1.18 (quin, 2H, J=7.62 Hz)	9.38 (s, 2H) 4.43 (t, 4H, J=7.05 Hz) 1.81 (quin, 4H, J=7.09 Hz) 1.22 (m, 6H)	9.39 (s, 2H) 4.43 (t, 4H, J=7.14 Hz) 1.81 (quin, 4H, J=7.09 Hz) 1.21 (m, 10H)
^{13}C-NMR (50 MHz, d^6 DMSO) [ppm]	143.8 (2C, d) 47.1 (2C, t) 28.3 (2C, t) 22.4 (2C, t)	143.7 (2C, d) 47.3 (2C, t) 28.9 (2C, t) 27.4 (t) 25.4 (2C, t)	143.7 (2C, d) 47.4 (2C, t) 29.0 (2C, t) 28.4 (t) 28.0 (2C, t) 25.5 (2C, t)
		Mid-FTIR cm^{-1}	
ν_s(C-H) aromatic,tetrazole ring	3115	3115	3115
ν_s(C-H), aliphatic spacer	2950 2870	2950 2870	2950 2849
ν_s(C-C), ν_s(C-N) tetrazole ring	1790 1490 1175	1792 1491 1174	1793 1492 -
elemental analysis calculated	C: 40.38 % H: 5.81 % N: 53.81 %	C: 45.75 % H: 6.83 % N: 47.42 %	C: 49.98 % H: 7.63 % N: 42.39%
found	C: 40.65 % H: 5.72 % N: 53.52 %	C: 45.97 % H: 6.90 % N: 47.32 %	C: 50.06 % H: 7.73 % N: 42.32%

2.2. SYNTHESIS AND CRYSTAL STRUCTURE OF THE LIGANDS

Table 2.2.: Crystallographic data of 5*ditz*, 7*ditz* and 9*ditz*.

	5*ditz*	7*ditz*	8*ditz*
Formula	$C_7H_{12}N_8$	$C_9H_{16}N_8$	$C_{11}H_{20}N_8$
F_w	208.25	236.30	264.35
cryst.size [mm]	0.60 x 0.22 x 0.20	1.00 x 0.25 x 0.04	1.00 x 0.25 x 0.04
Space group	$Fdd2$ (no. 43)	$Fdd2$ (no. 43)	$Fdd2$ (no. 43)
a [Å]	14.111(2)	13.738(2)	13.393(4)
b [Å]	32.041(4)	38.770(7)	45.260(13)
c [Å]	4.5613(6)	4.6300(8)	4.6724(14)
V,[Å^3]	2062.3(5)	2466.0(7)	2832.4(15)
Z	8	8	8
ρ_{calcd},gcm^{-3}	1.341	1.273	1.240
T,K	297(2)	297(2)	297(2)
μ [mm^{-1}](MoKα)	0.095	0.088	0.084
F(000)	880	1008	1136
θ_{max},[°]	25	30	25
no. of rflns meas	4273	8879	6658
no. of unique rflns	912	1788	1240
no. of rflns $I > 2\sigma(I)$	885	1365	984
no. of params	69	78	87
R_1 (I>2σ(I))[a]	0.0444	0.0391	0.0547
R_1 (all data)	0.0462	0.0558	0.0726
wR_2 (all data)	0.1083	0.1069	0.1549
Diff.Four.peaks, min/max,[eÅ^3]	-0.11/0.12	-0.10/0.12	-0.13/0.22

[a]$R_1 = \frac{\sum ||F_o|-|F_c||}{\sum |F_o|}$, $wR_2 = \left[\sum \left(w\left(F_0^2 - F_c^2\right)^2\right) / \sum \left(w\left(F_0^2\right)^2\right)\right]^{1/2}$

2.2. SYNTHESIS AND CRYSTAL STRUCTURE OF THE LIGANDS

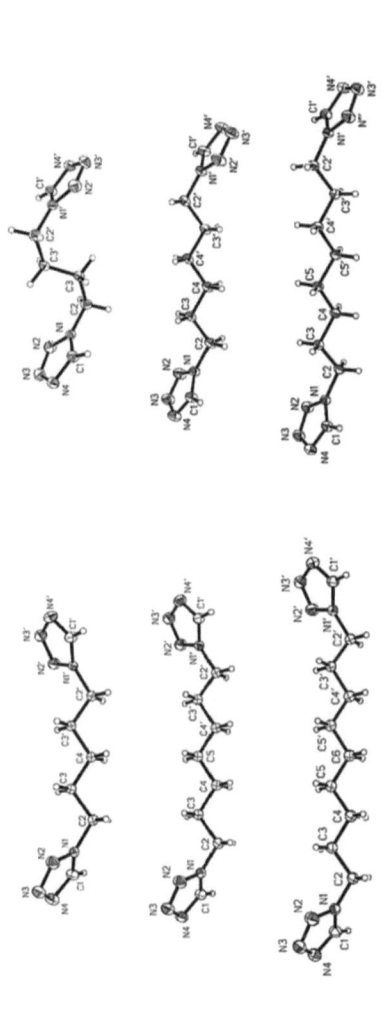

Figure 2.1.: Structural views of 5*ditz*, 7*ditz* and 9*ditz* (20% probability ellipsoids). All molecules have C_2 symmetry with the twofold axis passing through C4, C5 and C6, respectively. 4*ditz*, 6*ditz* and 8*ditz* are shown to the right for comparison [54]

2.3. Synthesis of the Complexes

The respective ligand (n*ditz*, 1mmol) was dissolved in hot reagant grade ethanol (n = 4-10 and 12). While the solution cooled down to 40 °C iron(II)tetrafluoroborate-hexahydrate or iron(II)perchlorate-hexahydrate (0.33 mmol) and a small amount of ascorbic acid to keep the iron as iron(II) were diluted in ethanol (5 ml). This solution was slowly added to the dissolved ligand and the resulting mixture stirred for four hours. The precipitate was filtered off and the obtained powder was dried over P_2O_5. Analytical data are presented in Table 2.3 and 2.4.

Synthesis of [Fe(4*ditz*)$_3$](X)$_2$ complexes with X = PF_6^-, ReO_4^-, SbF_6^- as well as [Fe(n*ditz*)$_3$](PF$_6$)$_2$ with n = 7-9 and [Fe(n*ditz*)$_3$](SbF$_6$)$_2$ with n = 7-9 can be found in [62] and [63]. Single crystals of [Fe(4*ditz*)$_3$](BF$_4$)$_2$ and [Fe(4*ditz*)$_3$](ClO$_4$)$_2$ were obtained by H-tube slow diffusion. 0.65 mmol ligand was dissolved in 10-15 ml hot solvent and placed in one side of the H-tube. On the other side of the tube 10-15 ml of an ethanolic solution containing 0.17 mmol iron(II)tetrafluoroborate-hexahydrate was added. For the ClO_4^- complex 0.17 mmol of iron(II)perchloratehexahydrate have been dissolved in $CHCl_3$.

The colorless single crystals of the complexes were obtained after 5 days ([Fe(4*ditz*)$_3$](ClO$_4$)$_2$) or fourteen days ([Fe(4*ditz*)$_3$](BF$_4$)$_2$) respectively.

Figure 2.2 and Figure 2.3 present the crystal structures of [Fe(4*ditz*)$_3$](BF$_4$)$_2$ and [Fe(4*ditz*)$_3$](ClO$_4$)$_2$. Discussion of the structures and crystallographic data are given in [62].

2.3. SYNTHESIS OF THE COMPLEXES

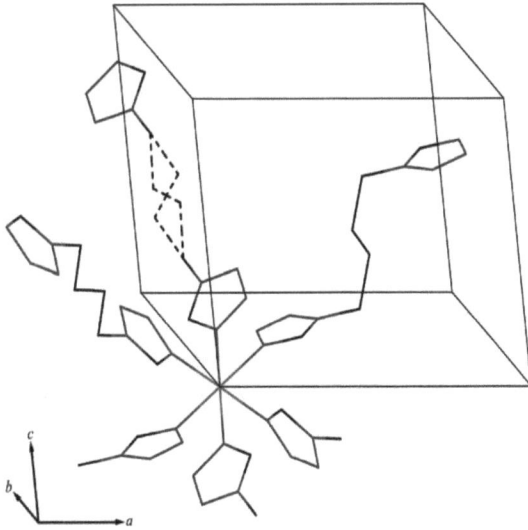

Figure 2.2.: Crystal structure of [Fe(4*ditz*)$_3$](BF$_4$)$_2$.EtOH at 89. Coordination environment of Fe(II). The disordered 4*ditz*-chain is drawn dashed, the unit cell is outlined. H atoms are omitted.

2.3. SYNTHESIS OF THE COMPLEXES

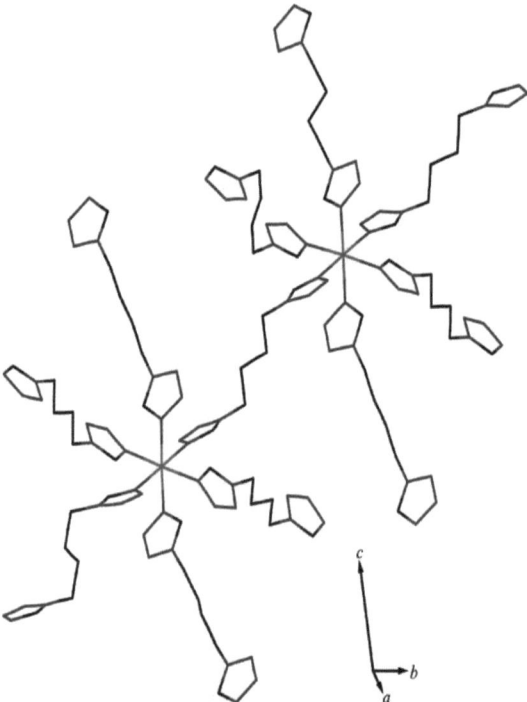

Figure 2.3.: Crystal structure of [Fe(4*ditz*)$_3$](ClO$_4$)$_2$.EtOH at RT. Coordination environment of the two iron sites. H atoms are omitted.

2.3. SYNTHESIS OF THE COMPLEXES

Table 2.3.: Yield, elemental analysis and mid-FTIR data of the [Fe(n*ditz*)$_3$](BF$_4$)$_2$ complexes.

	[Fe(n*ditz*)$_3$](BF$_4$)$_2$							
n	4	5	6	7	8	9	10	12
Yield [g, %]	0.20, 73	0.16, 59	0.15, 50	0.19, 61	0.18, 55	0.25, 72	0.24, 69	0.26, 67
Elemental Analysis								
calculated	C 26.62% H 3.72% N 41.40%	C 29.53 % H 4.25 % N 39.36 %	C 32.16 % H 4.72 % N 37.51 %	C 34.56 % H 5.16 % N 35.83 %	C 36.75 % H 5.55 % N 34.29 %	C 38.77 % H 5.91 % N 32.88 %	C 40.62 % H 6.25 % N 31.58 %	C 44.15 % H 6.35 % N 29.42 %
found	C 26.20 % H 3.75 % N 40.40 %	C 29.30 % H 4.18 % N 37.66 %	C 31.41 % H 4.78 % N 36.15 %	C 34.00 % H 5.18 % N 35.72 %	C 36.84 % H 5.58 % N 33.63 %	C 37.95 % H 5.83 % N 31.96 %	C 40.83 % H 6.03 % N 31.41 %	C 44.08 % H 6.90 % N 28.08 %
Mid-FTIR [cm^{-1}]								
ν_s(C-H), aromatic, tetrazole ring	3148	3146	3148	3147	3148	3148	3145	3141
ν_s(C-H), aliphatic spacer	2984 2958 2942	2949 2933 2862	2938 2862 -	2940 2862 -	2936 2859 -	2933 2857 -	2929 2851 -	2917 2852 -
$\nu_{N=N}$, ν_{C-N}	1507 1384	1506 1370	1506 1372	1507 1368	1507 1368	1507 1375	1504 1372	1501 1383
$\nu_{C=N}$, ν_{N-N}, ν_{C-N}	1182 1101	1183 1109	1182 1102	1173 1108	1183 1102	1174 1108	1177 1098	1176 1096

2.3. SYNTHESIS OF THE COMPLEXES

Table 2.4.: Yield, elemental analysis and mid-FTIR data of the [Fe(n*ditz*)$_3$](ClO$_4$)$_2$ complexes.

n	4	5	6	7	8	9	10	12
Yield [g, %]	0.22, 80	0.29, 82	0.21, 76	0.15, 43	0.32, 96	0.28, 86	0.20, 55	0.28, 71
				Elemental Analysis				
calculated	C 25.82% H 3.61% N 40.15%	C 28.68 % H 4.13 % N 38.23 %	C 31.28 % H 4.59 % N 36.48 %	C 33.66 % H 5.02 % N 34.69 %	C 35.83 % H 5.41 % N 33.43%	C 37.83 % H 5.77% N 32.08 %	C 39.68 % H 6.10 % N 30.85 %	C 43.19 % H 6.21 % N 28.78 %
found	C 26.59 % H 3.58 % N 39.36 %	C 29.65% H 4.03 % N 37.53 %	C 31.24 % H 4.45 % N 35.68 %	C 33.78 % H 4.48 % N 34.53%	C 36.64 % H 5.16 % N 32.88 %	C 38.930 % H 5.16 % N 32.88 %	C 39.03 % H 5.60 % N 29.23 %	C 42.57 % H 6.63 % N 28.01 %
				Mid-FTIR [cm^{-1}]				
ν_s(C-H), aromatic, tetrazole ring	3137	3135	3137	3137	3138	3138	3136	3139
ν_s(C-H), aliphatic butylene spacer	2956 2875	2947 2866	2935 2860	2940 2864	2934 2859	2929 2856	2917 2850	2922 2851
$\nu_{N=N}$, ν_{C-N}	1506 1368	1504 1363	1504 1360	1507 1372	1506 1364	1506 1366	1506 1373	1504 1374
$\nu_{C=N}$, ν_{N-N}, ν_{C-N}	1182 1086	1181 1097	1178 1093	1180 1089	1181 1093	1181 1094	1178 1089	1175 1094

2.4. Laboratory X-ray Powder Diffraction

The samples were gently ground and transferred to single crystal silicon sample holders by the slurry technique with cyclohexane (Merck, p.a.) as elutriating liquid. Powder patterns were recorded on a Philips X'Pert diffractometer in Bragg-Brentano geometry using Cu $K_{\alpha 1,2}$ radiation.

2.5. Synchrotron Powder Diffraction

Synchrotron radiation powder diffraction experiments were carried out by using the large Debye-Scherrer camera installed at the BL02B2 beamline, SPring-8 (Sayo-gun, Japan), that is equipped with an imaging plate detector [64]. With this camera, both high angular resolution and high counting statistics data can be collected. The as-precipitated phases were sealed in 0.3 mm glass capillaries. The X-ray powder patterns were measured from 300 K down to 9 K. (A He gas circulation type cryostat was used for the low temperature measurements.) All data were collected under the same experimental conditions except for the temperature. The exposure time of X-rays was 5 min. for each temperature. The wavelength of the incident X-rays was approx. 1 Å. The exact value was determined from a CeO_2 standard. The patterns were recorded with a step width of 0.01° in 2θ in the range $2\theta = 0\text{-}75°$, which corresponds to a resolution of $d = 0.82$ Å.

2.6. Scanning Electron Microscope (SEM)

SEM measurements have been made at the Aoyama Gakuin University on a JEOL JXA-8200 Electron Probe Microanalyzer and at the TU Wien on a JEOL JSM-T330A.

2.7. Optical Measurements

2.7.1. UV/VIS-NIR Measurements

UV/Vis–NIR spectra were recorded with a Perkin–Elmer Lambda 900 UV/Vis–NIR spectrometer between 1500 and 300 nm using the method of diffuse reflection. A spectrum of $BaSO_4$ was subtracted as background. Variable-temperature measurements were made using a custom-made thermostattable sample holder with quartz glass windows within a spectralon integration sphere. The temperature was controlled with a Harrick controller. Aluminium foil was used to improve the thermal contact between the sample holder and the sample. The spectra were measured between 100 and 303 K in intervals of 5 to 10 K.

2.7.2. Reflectivity Measurements

The reflectivity of the samples was investigated with a custombuilt reflectivity set-up equipped with a CVI spectrometer, which allows the collection of both the reflectivity spectra within the range of 450–950 nm at a given temperature and to follow the temperature dependence of the signal at a selected wavelength (± 2.5 nm) at 5–290 K. The analysis was performed on a thin layer of the powdered sample without any dispersion in a matrix [65]. A schematic illustration of the experimental setup is presented in Figure 2.4.

2.8. MAGNETIC MEASUREMENTS

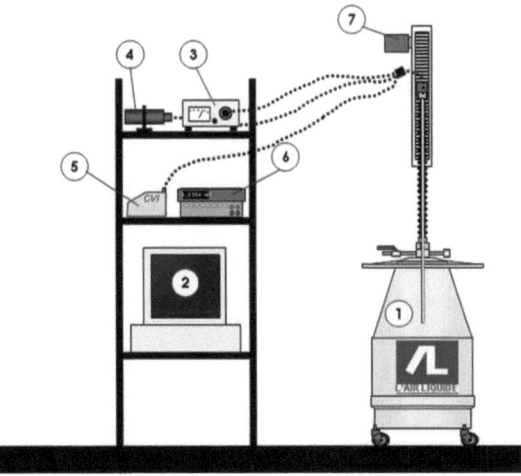

Figure 2.4.: Schematic view of the Reflectivity setup. With 1... helium dewar; 2... Control board; 3... lamp; 4... optical detector; 5... CVI spectrometer; 6... multimeter and 7... motorised sample holder

2.8. Magnetic and Magneto-optical Measurements

2.8.1. Magnetic measurements

Magnetic measurements were completed on three SQUID magnetometers:

(i) SQUID Cryogenix S600 magnetometer with an applied field of 1T (LAMM, Dipartimento di Chimica & UdR INSTM, Università di Firenze, Italy),

(ii) MPMS-55 Quantum Design SQUID magnetometer with an operating field of 2 T (Institut de Chimie de la Matière Condensée de Bordeaux, Université Bordeaux 1, France) and (iii) MPMSR2-RSO Quantum Design SQUID magnetometer with an operating field of 1 T within the temperature range of 2 - 300 K and with a cooling/heating rate of 10 Kmin^{-1} in the settle mode at atmospheric pressure. Further measurements were made on a 9 T-PPMS-system from Quantum Design VSM operating with a field of 1T. All magnetic measurements were performed on powder samples weighing ≈ 12 mg. The data

2.8. MAGNETIC MEASUREMENTS

were corrected for the magnetisation of the sample holder and for diamagnetic contributions, estimated from Pascal's constants.

2.8.2. Magneto-optical Measurements

The photo-magnetic measurements were performed with a Spectra Physics Series 2025 Kr$^+$ laser (λ=532 nm) coupled by an optical fibre to the cavity of the SQUID magnetometer (MPMS-55 Quantum Design SQUID) operating with an external magnetic field of 2 T within the 2–300 K temperature range and a speed of 10 Kmin^{-1} in the settle mode at atmospheric pressure. The quantity of the studied material is typically around 0.1 mg. The use of such small quantity is critical, particularly when the sample is couloured, because light penetration problems in bulk material should be avoided. The sample, in powder or crystal form, is deposited on commercial transparent adhesive tape placed close to the edge of an optical fiber installed in the rod sample holder, which is slowly placed down in the cavity of the SQUID. To eliminate oxygen the SQUID cavity is purged with gas, otherwise a magnetic peak is recorded in the 50 K region.

Another difficulty is to center the signal into the SQUID cavity. Because of the fact that both the LS state and the commercial transparent tape are diamagnetic, the resulting *emu* signal is negative, while the HS state is paramagnetic with a positive *emu* response. The magnetic signal between the initial and final states is then opposite; during the photo-excitation and/or relaxation process the signal irreversibly passes the zero *emu* position and any centering position is prohibited. Therefore the above mentioned reflectivity experiments are also used for the optimization of the light excitation. The power of the light excitation is adjusted to 5 mWcm^{-2}, which verifies that, when the light is switched off the magnetic response does not jump. Such an artificial effect is, in fact, a way to prove that a heating effect is happening on the sample. The weight of the thin layer samples is obtained by comparison of the measured thermal spin-crossover curve with another curve of a more accurately weighed sample of the same compound. The light irradiation is stopped when signal

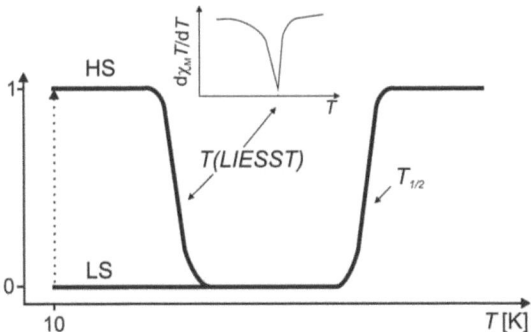

Figure 2.5.: Scheme and characteristic points of a LIESST experiment

saturation is reached *e.g.* when for a given laser power the equilibrium between the population and the relaxation is operating. This normally occurs after approximately one hour. After switching off the light (detection of the LIESST region) the time associated with three extractions used to detect the magnetic signal is 60 s and the time to reach the next temperature in steps of 1 K is 120 s. A characteristic procedure and the characteristic points of the experiment are illustrated in Figure 2.5.

2.9. Mössbauer Spectroscopy

The ^{57}Fe-Mössbauer spectra were recorded at selected temperatures within 4.2 - 294 K using a conventional constant acceleration drive system. The source used was ^{57}Co in a Rh-matrix with an activity of about 50 mCi. The data were analyzed using a least-squares fitting procedure assuming Lorentzian lines [66].

2.10. High Magnetic Field Measurements

2.10.1. Generation of Magnetic Fields

The Megagauss facility (Magnetotransport in Solids, Humboldt University at Berlin, Germany) can be used to produce magnetic fields up to 311 T [67]. For experiments this value is reduced to approx. 100 T. Fields of that magnitude have to be generated as pulsed magnetic fields [68, 69]. Therefore the technique of *single turn coils* (*stc*), which was introduced by Furth *et al.* [70], is used. Efficiency and simplicity of the experimental setup were demonstrated first by Forster and Martin [71]. In principle high electric electricity is created through fast discharge of capacitors and conducted in a lightweight single-use *stc*. The system comes up to an RCL-resonant circuit. Due to the magnetic pressure of over 40 GPa (10 tons cm^{-2}) in the megagauss range the coil disintegrates within microseconds. However, the disruption of the conductor does not affect the current flow which carries on by inductive plasma discharge. Since the outward bound radial acceleration of coil fragments does not harm the experimentally useful bore volume, the technique has been termed *semi-destructive*. Figure 2.6 gives an impression of the coil before and after the discharge.

Figure 2.6.: 12mm x 12mm x 3mm *scts* before the experiment (bottom), after a 10 kV, 6 kJ discharge with 37 T peak field (middle) and after a 55 kV, 189 kJ discharge with 188 T peak field (top).

2.10. HIGH MAGNETIC FIELD

The dimension of the magnetic field depends on the charging voltage of the capacitors as well as on the dimensions of the coil. Due to the fact that the charging voltage is limited to 60 kV, the magnetic field can only be increased by scaling down the coil dimensions. See Table 2.5 which gives the correlation between the coil dimensions and the reached fields.

Table 2.5.: Correlation between the coil dimensions and the induced magnetic fields.

coil[mm]	20×20×3	15×15×3	12×12×3	10×10×3	8×8×3	5×5×3
maximum field	114T	154T	188T	214T	261T	300T
rise time	2.7μs	2.5μs	2.3μs	2.0μs	1.7μs	1.3μs

The yield strength of copper permits peak fields of only 20 T for the non-destructive operation of the assumed standard coil. At higher fields the coil expansion is delayed only by mechanical inertia and the containment of magnetic flux must be supported by short rise times. Calculating the momentum transfer of a sinusoidal 200 T pulse on the standard coil yields a 2.4 μs rise-time requirement if the radial expansion is limited to 10 %. Although this condition is to some extent arbitrary, microsecond rise times are an intrinsic feature of *single-turn coils* and by far the most important technical boundary condition. All electrical parameters of the discharge circuit are fixed by the required rise time, a current of roughly 2.6 MA providing 200 T in the standard coil and technical feasibility. Assuming a thoroughly minimized total inductance of 20 nH and reasonable margins for resistive losses in a system with capacitive energy storage, the upper limits of 200 kJ and 120 μF for the energy and capacitance and a lower limit of 50 kV for the charging voltage are set. Following this guideline, the Berlin generator (see Figure 2.7) is composed of 20 capacitors, rated at 60 kV, 40 nH and 6 μF each, and 10 rail-gap switches, nominally 20 nH and a 750 kA, 1 C charge transfer limitation [72]. The short field rise time requires a rather precise ignition of high-voltage switches, as well as perfect synchronization of data recording equipment. The proper timing is achieved by a six channel, 0.1–10 μs delay trigger unit that controls both the recording systems and a 50/60 kV master/slave pulse generator with 10 ns rise time, a jitter of 1 ns [72], and separate output channels for each rail-gap switch. To avoid a disastrous accidental discharge of the entire capacitor bank via a single switch, the generator is composed of 10 separate modules, each including two capacitors and one switch in a high-voltage insulated housing. While switches are open the modules are only interconnected via 100 kΩ resistances, which are part of the charge and shutdown circuit. The high-voltage trigger–discharge operation produces tremendous electromagnetic perturbations and bears the risk of fatal accidents. All high-voltage units are therefore enclosed in an interlock secured Faraday cage. The remote control and data transfer lines entering the cage are exclusively based on optical fibres and pneumatic tubes to guarantee a complete galvanic separation and optimum shielding efficiency. Although in smaller proportions, stationary parts of the generator are exposed to the same type of electromagnetic forces as the single-turn coil. In the 1.0 m wide strip-line a 2.6 MA current produces a pressure of almost 1 MPa. However, the limited momentum transfer during the short pulse makes a rigid heavy weight support unnecessary. The Berlin generator is therefore mounted in a suspension frame with flexible shock absorbers stabilizing the large aluminium plates. For the monitoring of the inductive field probes and magnetization measurements inside the high-voltage area an electromagnetic pulse protected 8 bit, 200 MS s^{-1} digitizer with integrated optical converter was therefore devised [67].

2.10. HIGH MAGNETIC FIELD

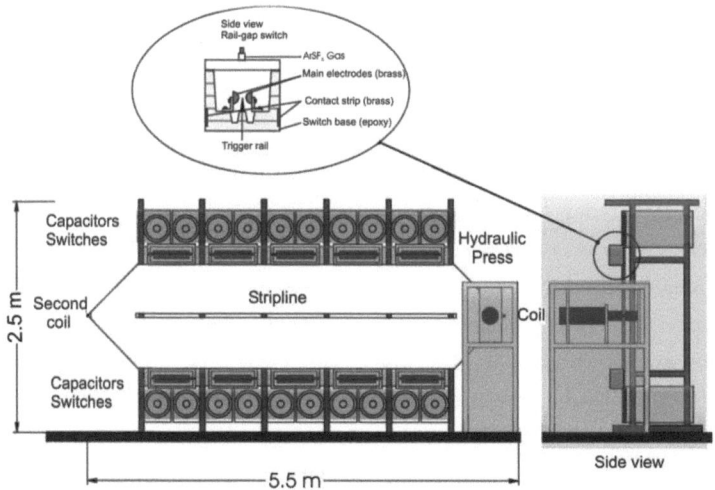

Figure 2.7.: Diagram of the Megagauss facility in Berlin

2.10.2. Detection of the Magnetic Field

The magnetic field is measured with an calibrated induction coil, the so called *pick up coil* (*puc*). The induced voltage U_{ind} is proportional to the derivative of the time of the trapped magnetic flux Φ (see equation 2.1).

$$U_{\text{ind}} = -\frac{d\Phi}{dt} = -\frac{d(\text{B} \cdot \text{A})}{dt} = -\frac{\text{A}d\text{B}}{dt} \quad (2.1)$$

A is the effective area of the *puc*. The *puc* consists of one turn, with a diameter of 2.5 mm and a 60 μ copper wire. Coils like that result in 1.5 kV induced voltage for a rising field of 300 T/μs.

Before saving and digitalisation the voltage is electronically integrated. A digital system, consiting of an integrator, digitiser, temporary storage, electro-optical converter and a battery-buffered current supply in a shielded box in the Fraraday cage, commutes the voltage into an optical signal. The sample rate is 200 MS/s. The error of the measurement of the field is less than 1 T in a

2.10. HIGH MAGNETIC FIELD

200 T signal. Induced magnetic fields during the measurements are presented in Figure 2.8 a.

2.10.3. Sample Holder and Cooling System

To detect the high-spin/low-spin transition a setup measuring the normal reflection in Faraday configuration using Plastic Optical Fiber (*POF*) for the monochromatic laser radiation of λ=632 nm or λ=541 nm is applied. The powder sample is fixed with a PE folio in the sample holder and is then carefully placed on the holder of the fibre until the *puc* contacts the surface of the sample. That warranted that the *puc* is as close as possible to the sample. Afterwards two *POF*s are fixed on the glass fiber bar. For a strong signal, the *POF*s must be aligned, so that a maximum of the reflected light is collected in the second *POF*. The thermocouple is also fixed close to the sample (see Figure 2.8 b) The sample holder was mounted in a miniature N_2-cryostat to meet the limited dimensions of the magnetic field coil. As detector a fast photo-diode with 125 MHz bandwidth was used.

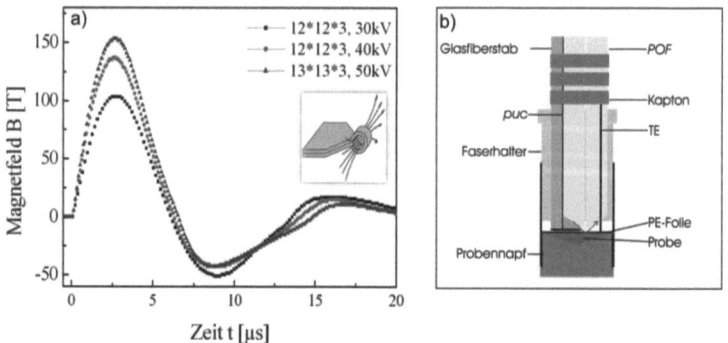

Figure 2.8.: a) The induced magnetic fields depend on the dimension of the coil and the charging voltage b) Profile of the justified system with the sample holder, *POF*, Thermal element (TE) and the light conductor system.

Results

3. Results

3.1. Powder Diffraction

The microcrystalline samples were investigated by XRPD. If one considers the visual appearance of the powder patterns the [Fe(n*ditz*)$_3$](BF$_4$)$_2$ series splits into two groups, namely the phase with the butylene spacer and the series with n = 5 - 10 and 12 (Figures 3.1 und 3.2). Incidentally, this division also holds for the corresponding perchlorate compounds [Fe(n*ditz*)$_3$](ClO$_4$)$_2$, whereupon these show a higher degree of crystallinity compared to the tetrafluoroborates: Whereas the patterns of the phases with $n >$ 4 are characterised by reflection broadening that increases with n, [Fe(4*ditz*)$_3$](X)$_2$ (X= BF$_4^-$, ClO$_4^-$) are well crystallised. During the synthesis of the BF$_4^-$ and ClO$_4^-$ phases with $n >$ 4 it turned out that the amount of water in the used ethanol influences the crystallinity of the powders. Therefore the powders were synthesised in EtOH / H$_2$O mixtures (abs. EtOH, 95% / 5% and 75% / 25%). The results of these series of experiments were on the one hand that the precipitation rate followed the order: abs. EtOH > 95% / 5% > 75% / 25% and on the other that the crystallinity of the powder was best when the used EtOH contained 5% water. The 75% / 25% mixture lead to co-precipitation of the ligand and the powders made in abs. EtOH showed in comparison with the others a higher amorphous content.

[Fe(4*ditz*)$_3$](BF$_4$)$_2$ exhibits a reflection distribution similar to its perchlorate analogue (Figure 3.1) as preiviously described [58]. There is, however, a mismatch in the reflection positions. Since van Koningsbruggen *et al.* [58] report a unit cell for the perchlorate an attempt was made to obtain lattice parameters for the tetrafluoroborate as well. The positions of the first 20 reflections were

3.1. POWDER DIFFRACTION

used for autoindexing using the ITO program [73] without prior knowledge of the unit cell later determined from a single crystal. The best solution was found in the triclinic crystal system, $a = 8.56$, $b=10.91$, $c=10.94$ Å, $\alpha=60.2$, $\beta=85.0$, $\gamma=90.0$ °, V=881 Å^3, $M_{20}=26$. The volume is in accordance with one formula unit of the complex per unit cell. As can be seen readily the cell shows pseudo-hexagonal symmetry with parameters lying closely to the trigonal phase [Fe(4$ditz$)$_3$](PF$_6$)$_2$.MeOH [52] (SG P-3, $a = 11.258(6)$, $c=8.948(6)$ Å, V= 982(1) Å^3, at 300 K. To allow better comparison the indexing solution was transformed by the matrix [0 1 0] [0 0 -1] [1 0 0] to a non-standard setting to resemble the lattice parameters of its parent PF$_6^-$ phase.). Comparative inspection of the powder patterns' relative intensities of the tetrafluoroborate and the hexafluorophosphate clarified that the BF$_4^-$ complex constitutes a distorted variant of the PF$_6^-$ salt. To back the result Le Bail [74] refinement was carried out to obtain more accurate lattice parameters utilising the program GSAS [75]. The first refinement trials with profile function no. 3 indicated pure microstrain broadening of the reflections ($\chi^2 = 26.8$). To improve the fit the anisotropic microstrain axis Y$_e$ [$h\,k\,l$] was permuted in the range -1 to 1 in order to obtain a better fit, that was found for the [0 1 1] axis ($\chi^2 = 20.7$). Since the agreement indices were quite bad at this stage it was switched to profile function no. 4 that includes a more sophisticated treatment of anisotropic microstrain broadening [76]. Indeed the refinement converged with R_{wp}=0.049 and χ^2=14.4 (see Figure 3.1). The quite large GoF results from the long counting time/step [77]. The analysis of the crystal structure of the [Fe(4$ditz$)$_3$](BF$_4$)$_2$ and [Fe(4$ditz$)$_3$](ClO$_4$)$_2$ complex can be found under [62]. The interpretation of the [Fe(4$ditz$)$_3$](ClO$_4$)$_2$.EtOH phase's pattern is less clear-cut. Le Bail fitting with the lattice parameters reported in [58], $a = 10.754(3)$, $b = 10.768(4)$, $c = 17.43(1)$ Å, $\alpha = 89.47(4)$, $\beta = 88.36(4)$, $\gamma = 60.46(3)°$, resulted in unacceptable agreement indices ($R_{wp}= 0.24$, $\chi^2 = 93$). The erroneous fit is readily apparent from the difference curve (Figure 3.1, top). Release of the lattice parameters produced unrealistic values. The reason for this are missing reflections and shifts in some angular reflection positions. In spite of these severe discrepan-

3.1. POWDER DIFFRACTION

cies it is obvious from Figure 3.1 that the perchlorate is structurally closely related to the BF_4^- and PF_6^- complexes. Due to the high reflection density, however, indexing of such a complex pattern is outside the scope of laboratory X-ray powder data. The apparent difference in the unit cells between the presented product and the phase reported earlier [58] is ascribed to the different solvents used to crystallise the complexes. Whereas van Koningsbruggen et al. [58] used methanol the present samples were precipitated from ethanol.

The homologous series [Fe(nditz)$_3$](X)$_2$.EtOH (X= BF_4^-, ClO_4^-), n = 5 - 10 and 12 is characterised as follows (Figure 3.2): i) The perchlorate compounds show a higher degree of crystallinity compared to the tetrafluoroborates.

ii) The diffractograms are very similar in shape. iii) The reflection with the longest d - spacing is the strongest and is shifted systematically to lower diffraction angles with increasing alkyl chain length. iv) The reflections broaden with increasing n and level off to background at $2\theta \approx 35°$, while the half-widths of the longest ones are more or less conserved. v) All phases form very thin plate-like microcrystallites with a thickness of a fraction of a micrometer. Point ii) suggests that the phases exhibit more or less the same basic structure. The reflection broadening can be explained by size effects due to the extremely thin plates. As the largest crystallographic axis is usually the direction of the lowest crystal growth rate, the base plane of the microplates lies perpendicular to the longest cell axis. The reflection with the largest d-spacing, showing a steady shift as n increases, can be regarded as this axis (possibly of higher order) that mirrors the increasing number of methylene groups in the ligands. Powder indexing of the phase [Fe(5ditz)$_3$](BF$_4$)$_2$ with the positions of the first 20 reflections using the DICVOL program [78] yielded a preliminary solution in the monoclinic crystal system (pseudo-hexagonal), a = 10.99, b = 20.37, c = 10.07 Å, $\alpha = \gamma = 90, \beta = 126.1°$, V = 1822 Å^3, M_{20} = 13, which is in accordance with a unit cell content of Z = 2. This metric can be regarded as a distorted variant of the trigonal symmetry found for [Fe(2ditz)$_3$](BF$_4$)$_2$ [56] with comparable lattice parameters, $a = b$ = 10.380(1), c = 14.953(3) Å, V = 1395.3(3) Å^3 (at 296(2) K, the shortened c-axis is caused by the ethylene spacer). Therefore

3.1. POWDER DIFFRACTION

we tend to describe the pentylene- to dodecylene-ditetrazole complex series in a chain-type arrangement as well. The higher order complexes are suspected to crystallise in the triclinic system which, however, is not possible to confirm from the low quality powder patterns. Calculating the unit cell volumes for the phases under investigation, assuming the monoclinic cell from above and setting the b-axis length to $d \times 2$ of the longest line (i.e. the 020 reflection), results in unit cell contents Z of about 2, which again is in accordance with the [Fe(2$ditz$)$_3$](BF$_4$)$_2$ complex.

3.1. POWDER DIFFRACTION

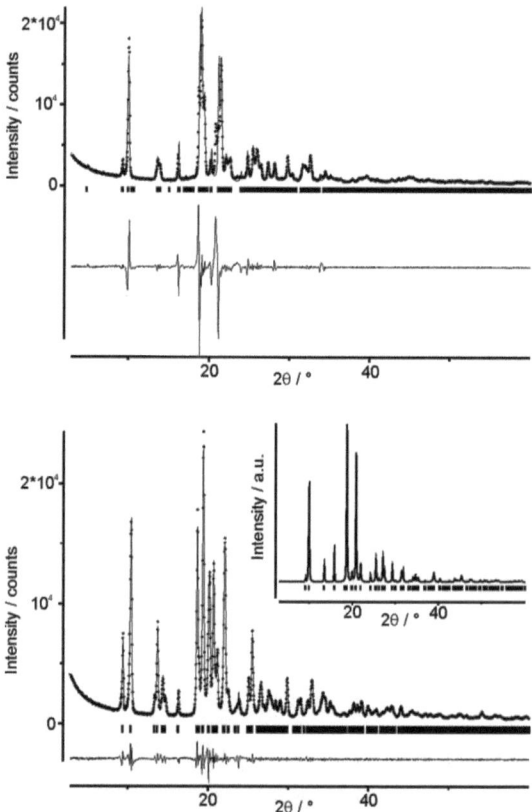

Figure 3.1.: Le Bail fitted X-ray powder patterns ($\lambda \ldots$ Cu K$_{\alpha 1,2}$) of the 4*ditz* phases with BF$_4^-$, bottom and ClO$_4^-$, top. Calculated profiles are drawn with solid lines, reflection positions are marked with vertical ticks and the difference curves are shown at the bottom. Inset is the calculated diffractogram of [Fe(4*ditz*)$_3$](PF$_6$)$_2$.MeOH.

3.1. POWDER DIFFRACTION

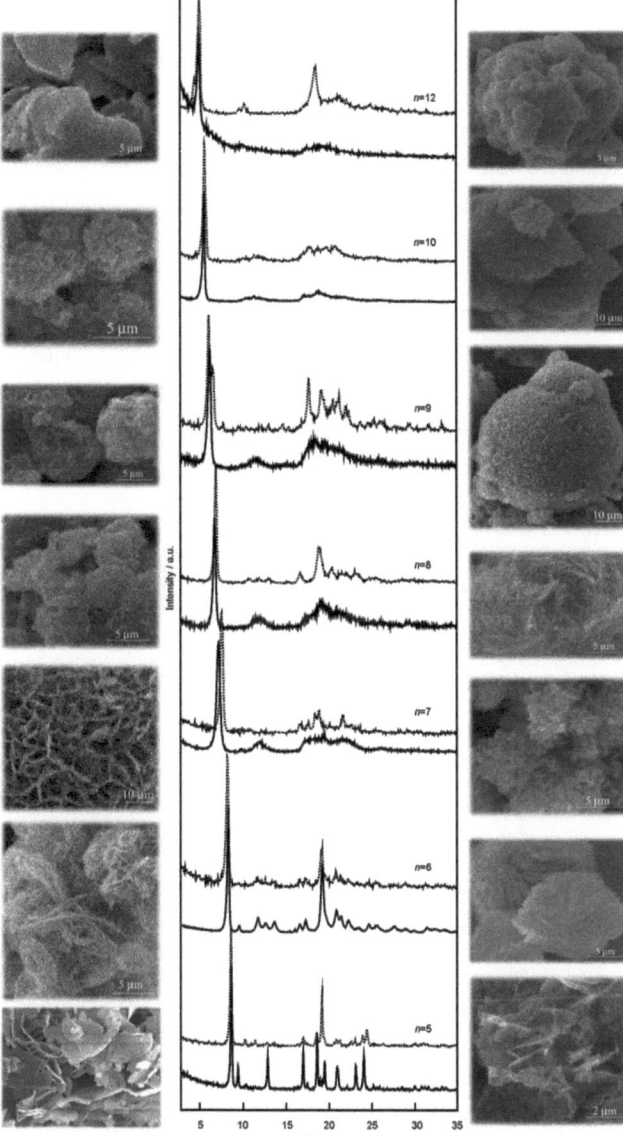

Figure 3.2.: Comparison of the [Fe(nditz)$_3$](BF$_4$)$_2$ (n = 5-10, 12, solid) series' X-ray powder patterns (λ ... Cu K$_{\alpha1,2}$) with the corresponding perchlorates (···), flanked by the corresponding SEM images (BF$_4^-$ left, ClO$_4^-$ right).

3.1. POWDER DIFFRACTION

3.1.1. Synchrotron Powder Diffraction

In order to get an overview of the temperature dependent shift of the reflections and to check for possible phase transitions well crystallised batches of [Fe(5$ditz$)$_3$](BF$_4$)$_2$, [Fe(8$ditz$)$_3$](BF$_4$)$_2$ and [Fe(7$ditz$)$_3$](ClO$_4$)$_2$ were measured at the beamline BL02B2 (wavelength approx. 1Å) installed at the synchrotron source SPring-8 (Hyogo, Japan). As this diffractometer is equipped with an imaging plate it is especially suited for rapid data collection. In Figure 3.3 the powder patterns of [Fe(5$ditz$)$_3$](BF$_4$)$_2$ at 300 K, 133 K, 50 K and 10 K are illustrated. Note that due to the longer exposure time (60 min.) higher intensities of the measurement at 10 K are observable in comparison to 5 min. scans. The change of the metal to ligand bondlength during the spin transition can be detected by the shift of the reflections. In the three mentioned compounds this shift is clearly visible see Figures 3.4 and 3.5.

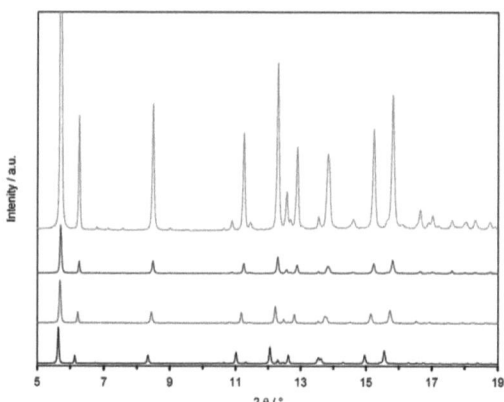

Figure 3.3.: Synchrotron powder patterns of [Fe(5$ditz$)$_3$](BF$_4$)$_2$ at 300 K (–), 133 K (–), 50 K (–) and 10 K (–) (note that the measurement time was 60 min. in comparison to 5 min. in all other measurements).

3.1. POWDER DIFFRACTION

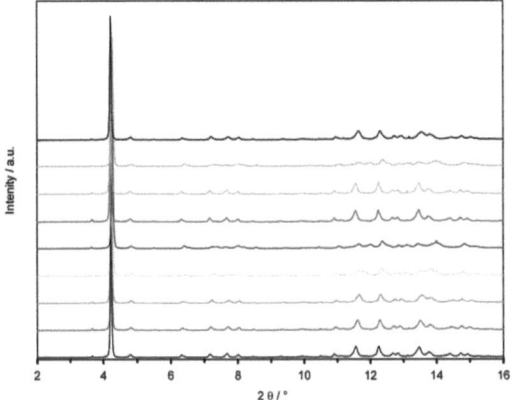

Figure 3.4.: Synchrotron powder patterns of [Fe(8*ditz*)$_3$](BF$_4$)$_2$ at 300 K (–), 250 K (–), 200 K (–), 150 K ( ), 100 K (–), 50 K (–), 9 K (–), 100 K (–) and 200 K (–).

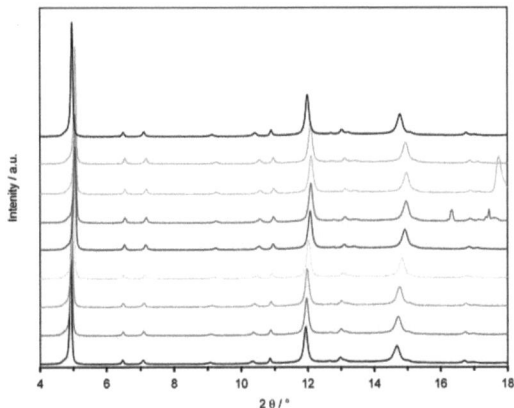

Figure 3.5.: Synchrotron powder patterns of [Fe(7*ditz*)$_3$](ClO$_4$)$_2$ at 300 K (–), 250 K (–), 200 K (–), 150 K ( ), 100 K (–), 50 K (–), 9 K (–), 100 K (–) and 200 K (–).

3.2. UV/VIS-NIR Spectroscopy

As expected all members of the series show a thermochromic effect associated with the spin transition, from white in the HS state and violet in the LS state. A typical example of a temperature - dependent spectrum, the one of [Fe(9$ditz$)$_3$](ClO$_4$)$_2$ is shown in Figure 3.6. The spectra are all recorded within the range 5000 - 35000 cm^{-1} (1500 - 300 nm) and at temperatures between 100 and 303 K. The absorption spectrum is composed of one charge-transfer (CT) band and three recognisable peaks, two of which decrease and one that increases with increasing temperature. The peaks at ≈ 18200 cm^{-1} and ≈ 26000 cm^{-1} are d-d transitions of the LS state of the complex, the peak at ≈ 33000 cm^{-1} is the CT band of the ligand and the band at ≈ 12000 cm^{-1} is the d-d transition of the HS state. According to the Tanabe-Sugano diagramm for d^6 systems, the two LS transitions can be assigned to the spin-allowed transitions $^1A_1 \rightarrow {}^1T_2$ and $^1A_1 \rightarrow {}^1T_1$ and the HS band corresponds to the $^5T_2 \rightarrow {}^5E$ transition.

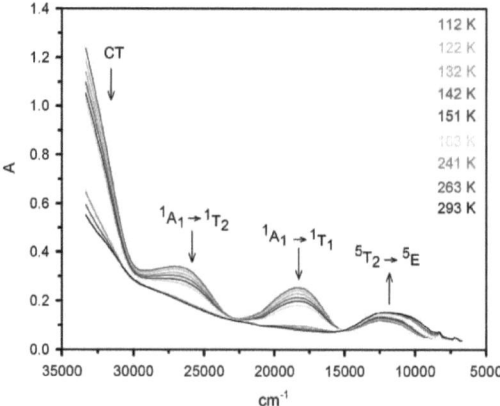

Figure 3.6.: Temperature-dependent UV/VIS-NIR spectra of [Fe(9$ditz$)$_3$](ClO$_4$)$_2$ between 5000 and 35000 cm^{-1} and the corresponding d-d transitions.

3.2. UV/VIS-NIR SPECTROSCOPY

3.2.1. Absorption spectra of the BF_4^- and ClO_4^- series

Due to the technical limitation of the cooling system (liquid nitrogen), in some complexes of the [Fe(n*ditz*)$_3$](BF$_4$)$_2$ and [Fe(n*ditz*)$_3$](ClO$_4$)$_2$ series (n = 4-9) the HS to LS transition is not complete. This means that the absorption bands of the complexes with very low $T_{1/2}$ ($\leq$ 150 K) are not strongly pronounced (see Figure 3.8 a,b,e and 3.9 b,c). If one looks in detail at the HS bands, it seems that in some compounds, except in the [Fe(6*ditz*)$_3$](BF$_4$)$_2$, [Fe(8*ditz*)$_3$](ClO$_4$)$_2$ and [Fe(9*ditz*)$_3$](ClO$_4$)$_2$, the HS band disappears not completely, respectively unsymmetrically. To verify this observation and to analyse the curves in more detail the following procedure was carried out: i) the area of the $^1A_1 \rightarrow {}^1T_1$ and $^5T_2 \rightarrow {}^5E$ transition is integrated and the calculated mole fraction is compared with the SQUID curves ii) the change of the HS band with temperature is fitted with a Gaussian curve iii) based on the energies of the HS and LS transitions, the ligand field splitting energy Δ_O is investigated.

The first step is to calculate the area of the $^1A_1 \rightarrow {}^1T_1$ and $^5T_2 \rightarrow {}^5E$ band. These values are converted into mole fraction (see equation 3.1 and Figure 3.5) to allow a comparison of the $T_{1/2}$ values with the more accurate magnetic measurement. As an example the result for [Fe(9*ditz*)$_3$](ClO$_4$)$_2$ is presented. The calculated $T_{1/2}$ value from the optical device is 170 K in comparison to 155 K from the magnetic data (see Figure 3.7). Subsequently the displayed temperature of the optical system is shifted by 15 K. This anomaly is representative for all performed experiments.

$$x = \frac{A - A_0}{A_{100} - A_0} \qquad (3.1)$$

x = molfraction
A = Area of the peak
A_0 = minimum area of the respective state
A_{100} = maximum area of the respective state

3.2. UV/VIS-NIR SPECTROSCOPY

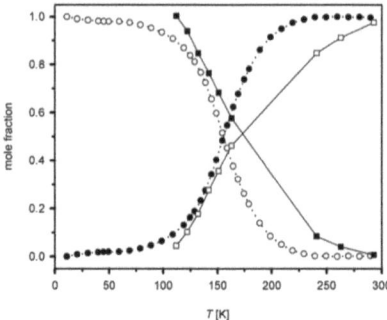

Figure 3.7.: Mole fraction of the optical measurement (squares) and magnetic measurement (dots) of [Fe(9$ditz$)$_3$](ClO$_4$)$_2$. The interception point corresponds to $T_{1/2}$.

As a result of this temperature shift it is self-evident that the motivation of the temperature dependent UV/VIS-NIR measurements is not the observation of the SC temperature but more the analysis of the obtained transitions of the complexes. If one looks in detail at the spectra, it is recognisable, that the appearance of the $^5T_2 \rightarrow {}^5E$ transition differs from the LS transitions. Hence this band is fitted with a Gaussian curve. The used parameters are given in equation 3.2.

$$A = A_{max} \exp\left(-(4\ln 2)\frac{\nu - \nu_{max}}{\Delta \nu}\right) \quad (3.2)$$

A = Absorption at ν
A_{max} = Band maximum
ν_{max} = wavenumber of the band
$\Delta \nu$ = half-width

Due to the fact that the left part of the HS bands between 12000 and 15000 cm^{-1} overlap with the LS transitions and that the HS band of Fe(II) seems to be consist of two absorptions bands, the band must be fitted with two or three Gaussian curves. During the attempts to fit the temperature depen-

3.2. UV/VIS-NIR SPECTROSCOPY

dent spectra, it was impossible to establish a logical sequence of bands and parameters which can be used for every HS band of the different compounds. Therefore it was decided to fit the band at RT respectively at low temperature (LT) with one Gaussian curve so that the change of the parameters can be compared between the complex series.

The first noticeable change of the HS bands is the shift of the maximum from RT to LT (RT → LT) which ends up at 600 cm^{-1} for [Fe(8*ditz*)$_3$](BF$_4$)$_2$ and 700 cm^{-1} for [Fe(8*ditz*)$_3$](ClO$_4$)$_2$ (see Figure 3.8 and 3.9 respectively the inserts). The displacement of the bands is reviewed in Table 3.1 for the [Fe(n*ditz*)$_3$](BF$_4$)$_2$ and [Fe(n*ditz*)$_3$](ClO$_4$)$_2$ (n = 4-9) series.

The comparison of the received values for the BF$_4$$^-$ and ClO$_4$$^-$ series (see Table 3.1) shows, that the shift for the same number of carbons in the spacer, independently of the anion, are in the same dimension. For $n = 8$ the highest values are observable in both series. This shift of the maximum of the $^5T_2 \rightarrow {}^5E$ transition is also responsible for the impression that only the right side of the curve disappears. But if one imagines that the area of the peak decreases at the same time as the band is shifted towards higher wavenumbers with temperature, it seems that the left side stays constant until the difference between these two effects is big enough that both changes are observable at the same time. This can be clearly seen in the [Fe(6*ditz*)$_3$](BF$_4$)$_2$ and [Fe(8*ditz*)$_3$](ClO$_4$)$_2$ complexes (see inserts Figure 3.8 c and Figure 3.9 e).

The most important point, regarding the $^5T_2 \rightarrow {}^5E$ transition, is the fact that the width of the HS transition also changes with temperature. The reason for that phenomenon is the following: if one acts on the assumption that in a Jahn-Teller distorted iron(II) complex (in solution [79]), the iron(II) HS band is composed of two absorption maxima (see 4.2) instead of one, it can be manifested that this also holds for the iron(II) in the discussed complexes. Because of the fact that the measurements are done on powder samples and measured in diffuse reflection, the separation of the two transition maxima is not possible. Instead of this one broad absorption band appears. Nevertheless, the variation of the

3.2. UV/VIS-NIR SPECTROSCOPY

width of the discussed HS band between RT and LT as well as between the different complexes should be an indication of the Jahn-Teller distortion of the complex. For comparison the widths of the HS transitions at RT for the BF_4^- and ClO_4^- series are presented in Table 3.2.

Table 3.1.: Summery of the displacement of the HS band maximum (RT $\rightarrow$ LT) for the [Fe(n$ditz$)$_3$](BF$_4$)$_2$ and [Fe(n$ditz$)$_3$](ClO$_4$)$_2$ (n = 4-9) complexes.

	Shift of the HS band maximum [cm^{-1}]	
	[Fe(n$ditz$)$_3$](BF$_4$)$_2$	[Fe(n$ditz$)$_3$](ClO$_4$)$_2$
n = 4	480	600
5	440	400
6	550	450
7	350	200
8	600	700
9	400	400

Table 3.2.: Widths of the $^5T_2 \rightarrow {}^5E$ transition at RT and LT for the [Fe(n$ditz$)$_3$](BF$_4$)$_2$ and [Fe(n$ditz$)$_3$](ClO$_4$)$_2$ (n = 4-9) complexes.

	[Fe(n$ditz$)$_3$](BF$_4$)$_2$		[Fe(n$ditz$)$_3$](ClO$_4$)$_2$	
Width [cm^{-1}]	RT	LT	RT	LT
n = 4	5500	4400	6990	5900
5	6200	5000	5000	4000
6	8100	5000	5500	4200
7	7000	5800	5000	4200
8	6000	4900	7000	7000
9	6500	3700	6000	5000

3.2. UV/VIS-NIR SPECTROSCOPY

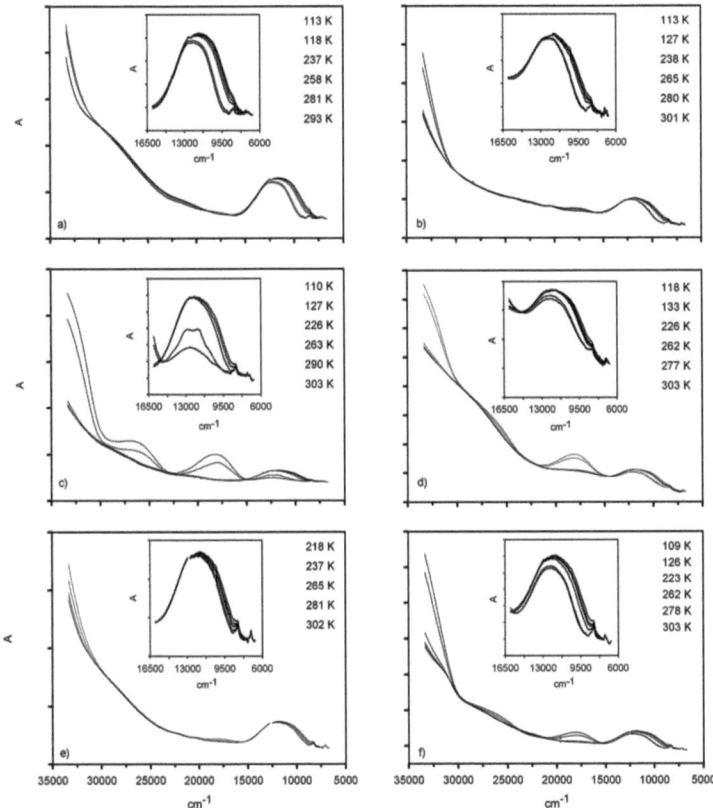

Figure 3.8.: Temperature dependent UV/VIS-NIR spectra of the BF_4^- complexes with n = 4-9 (a - f) in the range of 5000 - 35000 cm^{-1}. Inserts represent the $^5T_2 \rightarrow {}^5E$ transition.

3.2. UV/VIS-NIR SPECTROSCOPY

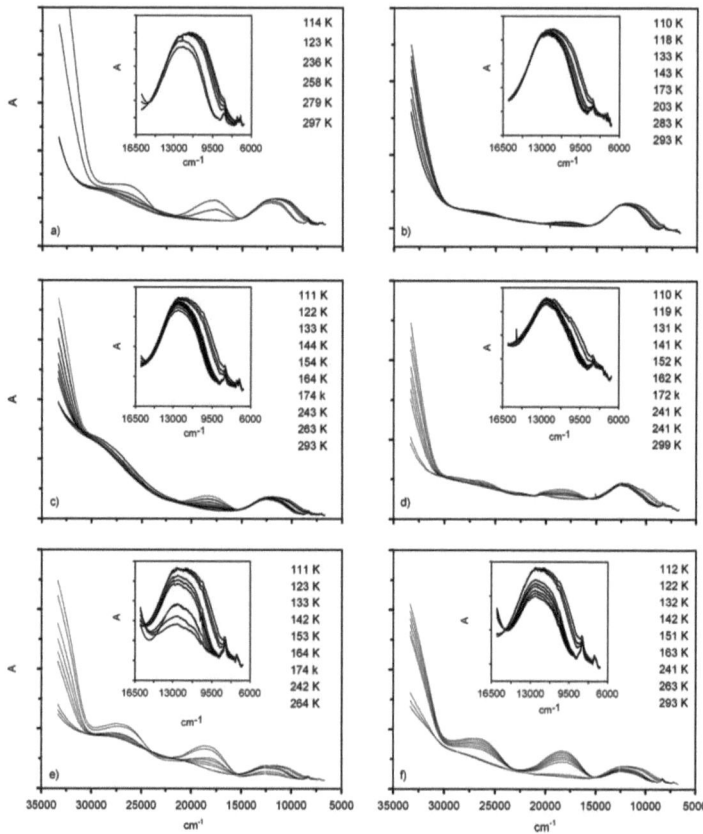

Figure 3.9.: Temperature dependent UV/VIS-NIR spectra of the ClO_4^- complexes with n = 4-9 (a - f) in the range of 5000 - 35000 cm^{-1}. Inserts represent the $^5T_2 \to {}^5E$ transition.

3.2. UV/VIS-NIR SPECTROSCOPY

Another important information which one can get out of the absorption spectra, is the Ligand field splitting energy Δ_O. There are two possibilities to obtain these values. The first one is the Tanabe-Sugano Diagramm for d^6 complexes, which shows the possible transitions as a function of Δ_O / B versus E/B where B represents the Racah-Parameter and E the Energy. The fact that the Tanabe-Sugano Diagramm allows only a graphical analysis of the Δ_O and B values of a certain complex, it is not the method which is suitable for the comparison of the two complex series discussed here. The differences between the complexes are too small for a graphical analysis. For that reason $\Delta_O(LS)$ is calculated from the energies of the transitions (see equation 3.3). Therefore the $^1A_1 \rightarrow {}^1T_1$ transition at LT is also fitted with a Gaussian curve. The overlap between the $^1A_1 \rightarrow {}^1T_2$ and the CT transition, makes fitting of this transition impossible. The energy of the $^1A_1 \rightarrow {}^1T_2$ transitions are therefore taken from the spectra.

$$\Delta_O(LS) = E(^1T_1) + \frac{E(^1T_2) - E(^1T_1)}{4} \qquad (3.3)$$

Δ_O for the HS transition is obtained from the maximum of the $^5T_2 \rightarrow {}^5E$ band [49]. The calculated values are summarized in Table 3.3.

Table 3.3.: Δ_O values for LS and HS for the BF_4^- and ClO_4^- complexes.

Δ_O	[Fe(n*ditz*)$_3$](BF$_4$)$_2$		[Fe(n*ditz*)$_3$](ClO$_4$)$_2$	
	LS	HS	LS	HS
n = 4	-^{a}	11610	20124	11500
5	19881^b	11830	20489	11800
6	20352	11750	20319	11900
7	20281	11800	20369	12150
8	20281	11500	20614	11900
9	20044	11800	20264	11800

^{a}No LS transition visible, $T_{1/2}$ outside the range of the instrument.
^{b} $^1A_1 \rightarrow {}^1T_1$ transition very small, no fit, E is read off the spectra.

3.2. UV/VIS-NIR SPECTROSCOPY

3.2.2. Influence of the Anions

As not only the influence of the carbons in the spacer on the absorption spectra is of interest, the following section focuses on the influence of the anion. Since the complexes with 4*ditz* as bridging ligands have been structurally analysed (see [62]) the temperature dependent UV/VIS-NIR spectra of the [Fe(4*ditz*)$_3$](BF$_4$)$_2$, [Fe(4*ditz*)$_3$](ClO$_4$)$_2$, [Fe(4*ditz*)$_3$](PF$_6$)$_2$, [Fe(4*ditz*)$_3$](ReO$_4$)$_2$ and [Fe(4*ditz*)$_3$](SbF$_6$)$_2$ are presented hereafter. If one compares these spectra with the already discussed ones, there are no great differences detectable. The energies of the $^5T_2 \rightarrow {}^5E$, $^1A_1 \rightarrow {}^1T_1$ and $^1A_1 \rightarrow {}^1T_2$ transitions are nearly the same as in the [Fe(n*ditz*)$_3$](BF$_4$)$_2$ and [Fe(n*ditz*)$_3$](ClO$_4$)$_2$ series. The biggest difference within the discussed complexes clearly appears in the HS transition of the [Fe(4*ditz*)$_3$](PF$_6$)$_2$ complex, which shows the only HS transition which disappears completely (see insert Figure 3.10 c). This can be ascribed to the high transition temperature of this compound. To underline the small differences between the spectra, the same systematically analysis is made as in the two series described before. The results are presented in Table 3.4. It can be seen that the Δ_O slightly increases with the size of the anion.

Table 3.4.: Displacement of the HS band maximum between RT and LT, width of the $^5T_2 \rightarrow {}^5E$ transition at RT and LT and Δ_O LS respectively Δ_O HS for the [Fe(4*ditz*)$_3$](X)$_2$ (X= BF$_4^-$, ClO$_4^-$, PF$_6^-$, SbF$_6^-$ and ReO$_4^-$) complexes.

	BF$_4^-$	ClO$_4^-$	PF$_6^-$	SbF$_6^-$	ReO$_4^-$
Shift HS band maximum [cm^{-1}]	480	600	900	550	500
Width $^5T_2 \rightarrow {}^5E$ transition at RT	5500	6200	5500	5500	7600
at LT	4400	5000	-^{a}	4400	6100
Δ_O LS [cm^{-1}]	-^{b}	20056^c	20056	20164	20314
Δ_O HS [cm^{-1}]	11610	11500	11700	11500	11500

^{a}HS transition disappears completely at LT.
^{b}No LS transition visible, T$_{1/2}$ out of the range of the instrument.
$^{c1}A_1 \rightarrow {}^1T_1$ transition very small, no fit, Energy is determined from the spectra.

3.2. UV/VIS-NIR SPECTROSCOPY

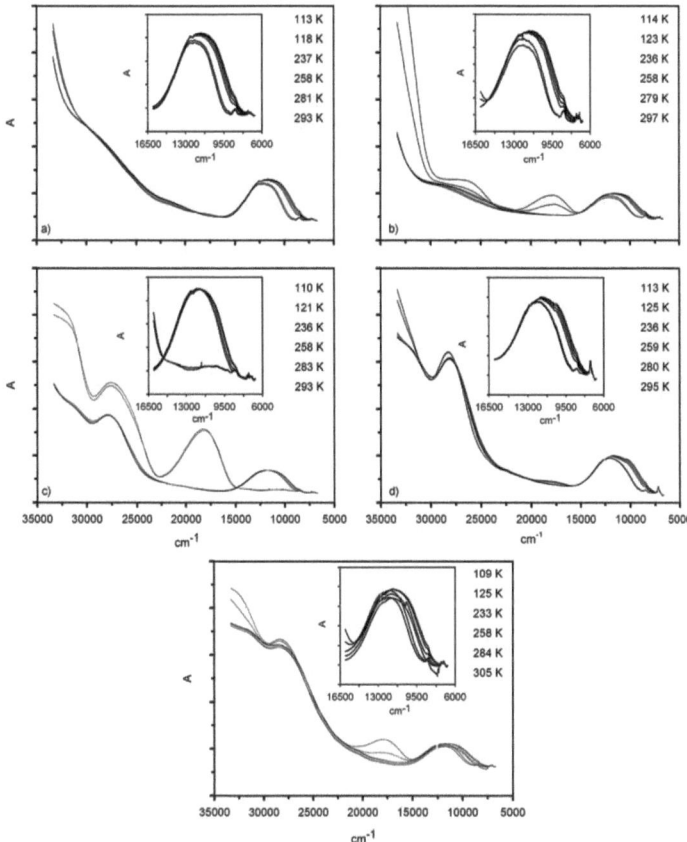

Figure 3.10.: Presentation of the temperature dependent UV/VIS-NIR spectra in the range of 35000 - 5000 cm^{-1} of (a) [Fe(4*ditz*)$_3$](BF$_4$)$_2$, (b) [Fe(4*ditz*)$_3$](ClO$_4$)$_2$, (c) [Fe(4*ditz*)$_3$](PF$_6$)$_2$, (d) [Fe(4*ditz*)$_3$](SbF$_6$)$_2$ and (e) [Fe(4*ditz*)$_3$](ReO$_4$)$_2$.

3.2.3. Solvent Effect

A very interesting topic concerning the discussed complexes is the influence of the solvent on the SC behaviour. Extensive discussion of this topic can be found in [62]. Hence a minimal example of changing the optical properties of the [Fe(4*ditz*)$_3$](ClO$_4$)$_2$.MeOH within the solvent amount is presented here. Two [Fe(4*ditz*)$_3$](ClO$_4$)$_2$.MeOH complexes were measured: one immediately after the synthesis (the powder was dried for approximately 15 min.) and the other after drying several days in the desiccator. The comparison of the LT measurements show, that the $^1A_1 \rightarrow {}^1T_2$ and $^1A_1 \rightarrow {}^1T_1$ transitions are not detectable in the dried complex but clearly visible in the solvated one (see Figure 3.11). This result leads to the conclusion that this complex needs a special amount of solvent to exhibit SC.

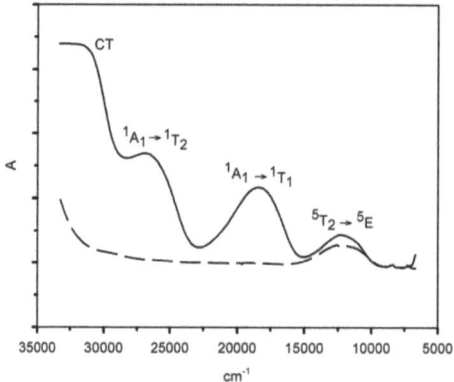

Figure 3.11.: Comparison of the [Fe(4*ditz*)$_3$](ClO$_4$)$_2$.MeOH "dry" (- -) and "wet" (–) complex measured at 101 and 120 K.

3.3. Reflectivity Measurements

In parallel to the temperature dependent UV/VIS-NIR measurements a custom-built reflectivity instrument to record the change occurring at 830±2.5 nm and at 550±2.5 nm, respectively the wavelengths of the $^5T_2 \rightarrow {}^5E$ and the $^1A_1 \rightarrow {}^1T_1$ transitions, as a function of temperature (290 - 10 K) and also light irradiation is used. The second LS transition, $^1A_1 \rightarrow {}^1T_2$, normally observable at 370 nm, is not within the range of the instrument. This experiment is done as a preliminary test for the LIESST experiment. Due to the fact that there is a great interest on compounds which show the LIESST effect the following series of complexes were measured: [Fe(n*ditz*)$_3$](BF$_4$)$_2$ with n = 4-9, [Fe(n*ditz*)$_3$](ClO$_4$)$_2$ with n = 4-9 and 12, [Fe(n*ditz*)$_3$](PF$_6$)$_2$ with n = 4, 7-9 and [Fe(n*ditz*)$_3$](SbF$_6$)$_2$ with n = 7-9. The goal was to get an idea how the chain length of the spacer and the size of the anion influences the LIESST properties of the compounds. A typical experiment is shown in Figure 3.12 for [Fe(5*ditz*)$_3$](ClO$_4$)$_2$ (spectra of all discussed complexes see appendix A).

3.3. REFLECTIVITY MEASUREMENTS

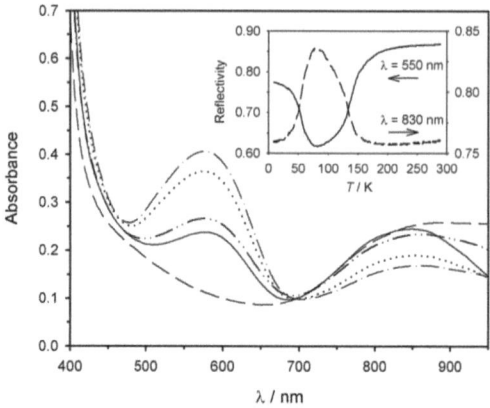

Figure 3.12.: Reflectivity measurements of [Fe(5ditz)₃](ClO₄)₂ as a function of temperature. In the main figure the absorbance spectra at (–) 280 K; (-··) 140 K, (-·-) 80 K, (···) 60 K and (–) 10 K is shown. The insert reports the reflectivity followed at 550±2.5 nm and 830±2.5 nm.

In agreement with the previous UV-VIS-NIR study, within the range of 450 - 950 nm at room temperature for all compounds only the HS band is visible whilst at 80 K only the LS transition is observable. Below 80 K, however, the reflectivity experiment demonstrates the existence of a photo-induced phenomenon at the surface. In fact, at low temperatures the sample was seen to bleach (see Figure 3.12 and insert) indicating the occurrence of a LS/HS photo-conversion through the LIESST effect. For each compound, the amount of photo-bleached fraction (% Irr-Surface; reported in Table 3.5) relative to the value found at room temperature was calculated. Interestingly, the level of photo-excitation roughly decreases with the length of the spacer in the [Fe(nditz)₃](ClO₄)₂ complexes. A similar trend could not be found for all the other series, but indeed, in agreement with the inverse energy-gap law [38], the higher the $T_{1/2}$, the lower the level of the light-induced LS/HS transformation. ($T_{1/2}$ is the temperature of the spin transition estimated from the maximum in

3.3. REFLECTIVITY MEASUREMENTS

the first derivative of the $\chi_M T$ vs. T plot.) Note that a value for [Fe(4*ditz*)$_3$](BF$_4$)$_2$ is less reliable than the others because $T_{1/2}$ is so low (approx. 80 K) that photobleaching occurs before a stable LS state is produced thus affecting the results. Furthermore, this sample does not switch more than 50% of its iron atoms to LS over all the considered temperature range.

Due to the fact that the discussed compounds (except [Fe(4*ditz*)$_3$](PF$_6$)$_2$) show the LIESST phenomenon on the surface, this effect seems to appear independent of the size of the anion, respectively the spacer length. Another noticeable fact is the appearance of a light induced thermal hysteresis (LITH) in almost all complexes. The best example for such an effect is found in the [Fe(4*ditz*)$_3$](BF$_4$)$_2$. The changes in the HS band intensity during light irradiation while cycling the temperature are presented in Figure 3.13. As the system is cooled, the reflectivity at 830 nm reaches a maximum at approximately 70 K and below this temperature bleaching occurs. As the sample is heated again the maximum occurs at approximately 80 K. This seems to indicate the presence of a thermal hysteresis during the spin transition which is confirmed by the SQUID measurements described hereafter. The existence of an hysteresis associated with the LIESST phenomenon corresponds to the LITH. This means that in this complex both phenomena, namely the LITH and the thermal hysteresis are detectable.

3.3. REFLECTIVITY MEASUREMENTS

Table 3.5.: % Irradiation surface caculated from the reflectivity measurements of the discussed complexes

[Fe(n*ditz*)$_3$](X)$_2$	X = BF$_4^-$	ClO$_4^-$	PF$_6^-$	SbF$_6^-$
	% Irradiation surface[a]			
4	52[b]	52	-	-
5	71	64	-	-
6	34	42	-	-
7	47	42	17	7
8	47	14	40	26
9	50	26	54	41
10	-	-	-	-
12	-	28	-	-

[a] Percentage of the sample photo-bleached
[b] This calculation is affected because the close vicinity of $T_{1/2}$ and *T(LIESST)* causes photo-bleaching to occur before a stable LS is reached

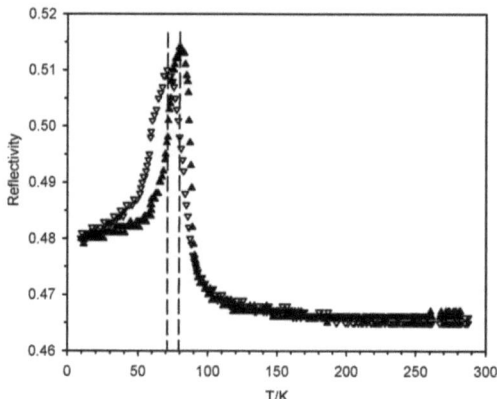

Figure 3.13.: Reflectivity of [Fe(4*ditz*)$_3$](BF$_4$)$_2$ followed at 830±2.5 nm as a function of temperature. Cooling (▽); warming (▲).

Another interesting result of the reflectivity measurement is the fact that

3.3. REFLECTIVITY MEASUREMENTS

the already well studied [Fe(4*ditz*)$_3$](PF$_6$)$_2$ complex, shows no LIESST effect (see Figure 3.14).

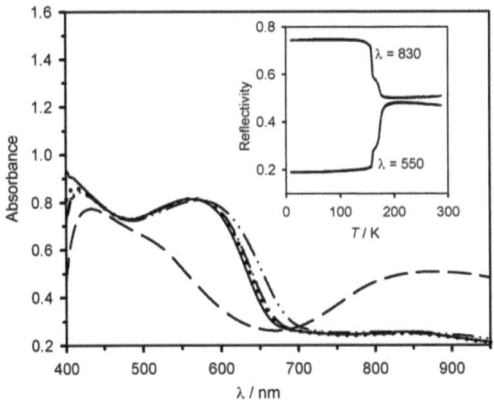

Figure 3.14.: Reflectivity measurement of [Fe(4*ditz*)$_3$](PF$_6$)$_2$.

3.4. Far-FTIR properties of the ClO_4^- series

Far-FTIR spectra of all the members of the $[Fe(nditz)_3](ClO_4)_2$ series were recorded. Interestingly almost no difference was observed (see Table 3.6). Unfortunately it was impossible to analyse the far-FTIR spectra of the $[Fe(nditz)_3](BF_4)_2$ complexes.

Table 3.6.: Temperature dependent far IR bands that can be associated with HS and LS species of $[Fe(nditz)_3](ClO_4)_2$.

n	HS	LS
5	469 (w); 388(w); 367(m)	421(s); 380(w); 358(sh); 287(s)
6	476(m); 378(s); 357(sh); 282(sh)	422(m); 393(s); 364(m)
8	437(sh); 373(m); 338(m); 310(sh)	424(s); 395(w); 379(w); 354(s)
9	366(sh); 331(sh)	422(s); 354(m)

3.5. Magnetic Measurements

In order to study how the number of carbon atoms in the spacer of the ligand and the size of the anion influences the spin transition, susceptibility curves were recorded between 10 and 300 K for the complexes of the [Fe(n*ditz*)$_3$](ClO$_4$)$_2$ and [Fe(n*ditz*)$_3$](BF$_4$)$_2$ series with n = 4-10 and 12. The obtained curves reporting $\chi_M T$, where χ_M is the molar magnetic susceptibility and T is the temperature, versus T are shown in Figure 3.15, Figure 3.16 and Figure 3.17. The temperature of the thermal spin state transition, $T_{1/2}$, estimated from the maximum in the first derivative of the $\chi_M T$ vs. T plot is given in Table 3.7.

3.5.1. Magnetic properties of the [Fe(n*ditz*)$_3$](ClO$_4$)$_2$ series

The investigated products of the [Fe(n*ditz*)$_3$](ClO$_4$)$_2$ series with n = 4-10 and 12 undergo a thermal spin transition at around 150 K, from a $\chi_M T$ product close to 3.0 cm^{-3}Kmol^{-1} at room temperature in agreement with the expected HS state, and a diamagnetic value at low temperature reflecting the LS state. At low temperatures in some of the curves a residual magnetization is observed yielding a $\chi_M T$ value up to 1 cm^{-3}Kmol^{-1}, which might be caused by traces of iron(III), thermally SC inactive HS iron(II) or zero-field splitting (ZFS).

Interestingly, an increase in the length of the spacer raises the $T_{1/2}$ value, but it can also be seen that the influence of the parity of the spacer is not negligible. Indeed, it seems that depending on the parity of the spacer, the complexes can be divided into two series: the first having a bridging ligand with odd-numbered carbon atoms in the spacer, which shows a $\chi_M T$ product at room temperature always equal to 3.0 cm^{-3}Kmol^{-1}; and a second series with even-numbered spacers, which exhibits a $\chi_M T$ at room temperature of approximately 3.75 cm^{-3}Kmol^{-1}. A similar classification can also be made if the shape of the thermal spin transitions is compared in (see Figure 3.15). Complexes with an odd-numbered spacer ([Fe(5*ditz*)$_3$](ClO$_4$)$_2$, [Fe(7*ditz*)$_3$](ClO$_4$)$_2$ and [Fe(9*ditz*)$_3$](ClO$_4$)$_2$) display a more gradual transition than complexes with an even-numbered ligand ([Fe(4*ditz*)$_3$](ClO$_4$)$_2$,

3.5. MAGNETIC MEASUREMENTS

[Fe(6$ditz$)$_3$](ClO$_4$)$_2$ and [Fe(8$ditz$)$_3$](ClO$_4$)$_2$). However, within such a classification the case of [Fe(4$ditz$)$_3$](ClO$_4$)$_2$ is somewhat exceptional, in that it is the only one displaying a two-step thermal spin transition with a plateau between 93 K and 124 K (Figure 3.15b).

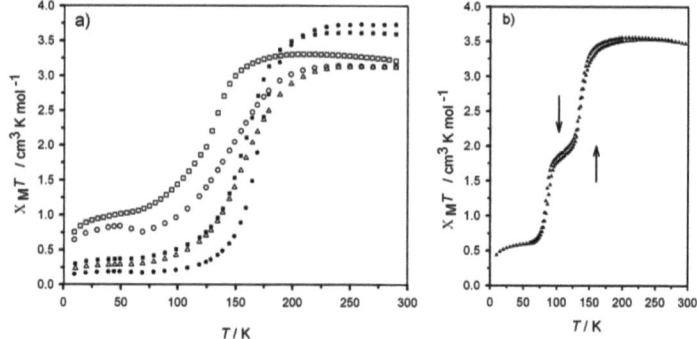

Figure 3.15.: Temperature dependency of the $\chi_M T$ product for the investigated complexes (▲) [Fe(4$ditz$)$_3$](ClO$_4$)$_2$, (□) [Fe(5$ditz$)$_3$](ClO$_4$)$_2$, (■) [Fe(6$ditz$)$_3$](ClO$_4$)$_2$, (○) [Fe(7$ditz$)$_3$](ClO$_4$)$_2$, (●) [Fe(8$ditz$)$_3$](ClO$_4$)$_2$ and (△) [Fe(9$ditz$)$_3$](ClO$_4$)$_2$ b) cooling and heating modes for [Fe(4$ditz$)$_3$](ClO$_4$)$_2$.

Table 3.7: Spin transition and LIESST parameters of the discussed complexes.

	X = BF_4^-			ClO_4^-		
[Fe(ndit$z$)$_3$](X)$_2$	$T_{1/2}$ [K]	T(LIESST)	% Irr-Surface[a]	$T_{1/2}$	T(LIESST)	% Irr-Surface
n = 4	80/89[b]	63	79[c]	84/134[d]	58/39	60
5	131	58	84	125	52	72
6	164	-[e]	25	155	-	43
7	154	58	71	144	51	57
8	148	52	55	169	-	13
9	154	50[f]	36	155	37	47
10[g]	150	-	-	160	-	-
12[h]	156	-	-	159	-	-

[a] Percentage conversion to the metastable high spin state by irradiation at 10 K
[b] Determined as the maximum of the first derivative of the cooling and warming curve
[c] 10 K the sample remains in a mixed LS-HS state, thus this refers to the % of "switchable" ions
[d] Determined as the maximum of the first derivative of the two transitions
[e] No minimum in $\delta\chi_M T/\delta T$ observable
[f] No minimum in $\delta\chi_M T/\delta T$ observable - estimated from the obtained relaxation curve of the LIESST experiment
[g] No magneto-optical data available
[h] No magneto-optical data available

3.5. MAGNETIC MEASUREMENTS

3.5.2. Magnetic Properties of the [Fe(n*ditz*)$_3$](BF$_4$)$_2$ Series

The compounds of the [Fe(n*ditz*)$_3$](BF$_4$)$_2$ series with n = 4-10 and 12 undergo, similar to the compounds of the [Fe(n*ditz*)$_3$](ClO$_4$)$_2$ series, a complete spin transition between 100 and 220 K (see Figure 3.16 and 3.17) and reach a $\chi_M T$ product close to 3.3 cm^{-3}Kmol^{-1}. The residual magnetisation up to 0.5 cm^{-3}Kmol^{-1} at low temperatures can be attributed to the same reasons as mentioned under 3.5.1. Comparison of the magnetic data with the perchlorate data shows a similar trend from n = 5-7 (see Table 3.7). The temperatures of transition, $T_{1/2}$, are all higher for BF$_4^-$ than for ClO$_4^-$. The SC curves themselves are also very similar to the [Fe(5*ditz*)$_3$](BF$_4$)$_2$ transition curve but a little steeper. There is, however, a disjunction when n is changed from 7 to 8. Here $T_{1/2}$ of [Fe(8*ditz*)$_3$](BF$_4$)$_2$ is lower and the transition itself is a lot more gradual than its perchlorate analogue. By n = 9, the transition is very gradual and almost exactly the same in both cases ($T_{1/2}$ = 154 K for BF$_4^-$ and 155 K for ClO$_4^-$). [Fe(10*ditz*)$_3$](ClO$_4$)$_2$ has a flatter magnetic curve than the BF$_4^-$ analogue but $T_{1/2}$ is 10 K higher (see Figure 3.17). But by n = 12, the transitions are remarkably similar again (see Figure 3.17).

The exception of the series is again the complex with 4 carbons in the spacer, namely the [Fe(4*ditz*)$_3$](BF$_4$)$_2$(see Figure 3.16). It shows an incomplete SC of approximately 50 % of the iron centres with an hysteresis of 9 K ($T_{1/2}$ of cooling = 80 K, warming = 89 K). Below 25 K a further sharp decrease of $\chi_M T$ is observed. This drop might be associated with a second spin transition at around 5 K. This would mean a shift of both SC temperatures by some 50 - 60 K. However this sharp decrease can also be associated with ZFS.

3.5. MAGNETIC MEASUREMENTS

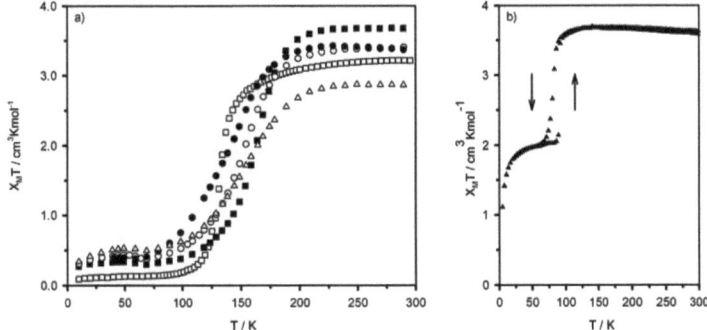

Figure 3.16.: Temperature dependency of the $\chi_M T$ product for the investigated complexes, a) ($\square$) [Fe(5*ditz*)$_3$](BF$_4$)$_2$, ($\blacksquare$) [Fe(6*ditz*)$_3$](BF$_4$)$_2$, ($\circ$) [Fe(7*ditz*)$_3$](BF$_4$)$_2$, ($\bullet$) [Fe(8*ditz*)$_3$](BF$_4$)$_2$ and ($\triangle$) [Fe(9*ditz*)$_3$](BF$_4$)$_2$ b) cooling and heating modes for [Fe(4*ditz*)$_3$](BF$_4$)$_2$.

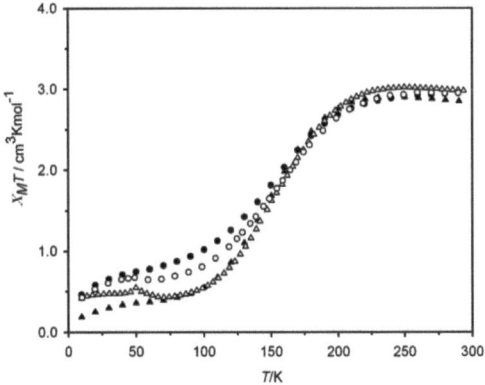

Figure 3.17.: Temperature dependence of $\chi_M T$ for (a) ($\blacktriangle$) [Fe(10*ditz*)$_3$](BF$_4$)$_2$ and ($\bullet$) [Fe(10*ditz*)$_3$](ClO$_4$)$_2$ b)($\triangle$) [Fe(12*ditz*)$_3$](BF$_4$)$_2$ and ($\circ$) [Fe(12*ditz*)$_3$](ClO$_4$)$_2$ in the heating mode. The existence of a small anomaly at around 50K on the curves almost certainly corresponds to an oxygen contamination.

3.6. Photo-Magnetic Measurements

The photo-magnetic properties of the different compounds have been investigated by following the effect of irradiation with light under the influence of an applied magnetic field. At first the sample was cooled slowly down to 10 K in order to stabilise the low spin state; then it was irradiated and the change in magnetism was followed (see Figure 2.5). When the saturation point had been reached the light was switched off, the temperature increased at a rate of 0.3 K min^{-1} and the magnetisation measured every 1 K. This procedure allows the quantification of $T(LIESST)$, which is determined by the minimum of the $\delta\chi_M T/\delta T$ vs. T curve recorded during relaxation [80].

3.6.1. LIESST Properties of the ClO_4^- Complexes

Figure 3.18 shows the photo-magnetic behaviour together with - for the sake of completeness - the magnetic susceptibility for each compound of the $[Fe(nditz)_3](ClO_4)_2$ with n = 4-9 series. Table 3.7 collects for each compound the highest percentage of photo-conversion obtained by irradiation at 10 K relative to the magnetic value recorded at room temperature for a pure HS state (%Irr-Bulk). As expected from the reflectivity analysis, all compounds undergo a light induced spin transition. If the shape of the $T(LIESST)$ curve recorded by increasing the temperature is compared, the complexes continue to behave differently one from the other. For those having a short spacer, the $\chi_M T$ product increases in the 10-30 K range while for long spacers the magnetic signal strongly decreases. This increase of the magnetic response with the temperature can be attributed to the effect of the zero-field splitting (ZFS) of the iron(II) HS in non perfectly octahedral geometry [81]. Indeed, the series of odd-numbered ligands shows a ZFS effect that progressively vanishes in the order $[Fe(5ditz)_3](ClO_4)_2$ > $[Fe(7ditz)_3](ClO_4)_2$ > $[Fe(9ditz)_3](ClO_4)_2$, reflecting a decrease of the lifetime of the photo-induced HS state in the tunneling region. An additional way to see the stability of the photo-induced HS state is to compare the magnitude of the $T(LIESST)$ temperatures [80, 40]. Table 3.7 collects

3.6. PHOTO-MAGNETIC MEASUREMENTS

$T(LIESST)$ temperatures of the [Fe(n$ditz$)$_3$](ClO$_4$)$_2$ family, with the exception of the [Fe(6$ditz$)$_3$](ClO$_4$)$_2$ and [Fe(8$ditz$)$_3$](ClO$_4$)$_2$ complexes where the minimum on the $\delta\chi_M T/\delta T$ vs. T curve can not be properly determined due to the very low efficiency of the photo-excitation (Figure 3.18). Nevertheless, it is interesting to see that the lowest $T(LIESST)$ temperature is found for [Fe(9$ditz$)$_3$](ClO$_4$)$_2$ which presents the highest $T_{1/2}$ value. Finally, it is important to note that the peculiarity of [Fe(4$ditz$)$_3$](ClO$_4$)$_2$, which is characterized by a two-step thermal spin transition, shows similar features in the $T(LIESST)$ experiments. Analysis of the $\delta\chi_M T/\delta T$ vs. T curve clearly shows the existence of two minima, at 39 K and 58 K, proving that both magnetically non-equivalent iron(II) metal centres can be photo-excited. It is also reasonable to propose that the $T(LIESST)$ temperature found at 58 K certainly corresponds to photo-excitation of the iron(II) metal centres involved in the LS/HS thermal spin transition occurring at 84 K, and the $T(LIESST)$ temperature at 39 K is linked to SC phenomenon occurring at 134 K.

3.6. PHOTO-MAGNETIC MEASUREMENTS

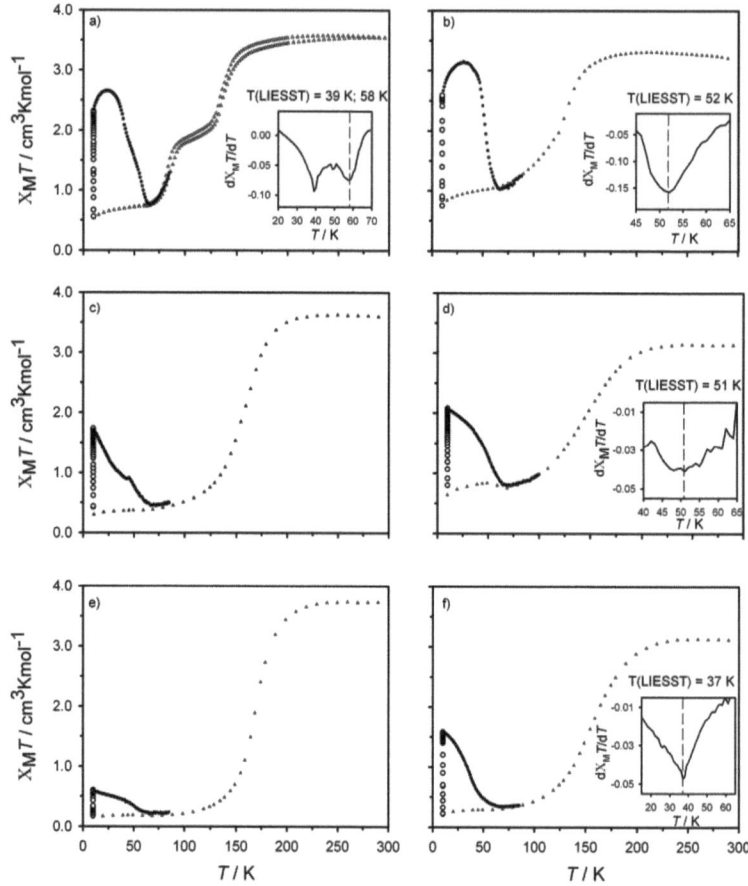

Figure 3.18.: Temperature dependence of $\chi_M T$ for (a) [Fe(4$ditz$)$_3$](ClO$_4$)$_2$, (b) [Fe(5$ditz$)$_3$](ClO$_4$)$_2$, (c) [Fe(6$ditz$)$_3$](ClO$_4$)$_2$, (d) [Fe(7$ditz$)$_3$](ClO$_4$)$_2$, (e) [Fe(8$ditz$)$_3$](ClO$_4$)$_2$, (f) [Fe(9$ditz$)$_3$](ClO$_4$)$_2$. (△) Data recorded without irradiation; (○) data recorded with irradiation at 10 K; (●) T (LIESST) measurement, data recorded in the warming mode with the laser turned off after irradiation for one hour. (Note that the small bump at 49 K observable in (c) is due to a small amount of oxygen).

3.6.2. LIESST Properties of the BF_4^- Series

The results of the LIESST experiments for the [Fe(n*ditz*)$_3$](BF$_4$)$_2$ with n = 4-9 complexes are shown in Figure 3.19 together with the magnetic susceptibility for each compound and Table 3.7 collects the %Irr-Bulk values. At first view there are no significant differences visible between the complexes with ligands yielding more than 4 carbons in the spacer in comparison to the results of the [Fe(n*ditz*)$_3$](ClO$_4$)$_2$ series. [Fe(5*ditz*)$_3$](BF$_4$)$_2$ shows like the ClO$_4^-$ analogon an increase of $\chi_M T$ after irradiation which can be attributed to ZFS. In the case of the [Fe(6*ditz*)$_3$](ClO$_4$)$_2$ the change of the anion leads to a shift of the $T_{1/2}$ to higher temperatures and consequently the LIESST effect is less pronounced. The change of the anion from ClO$_4^-$ to BF$_4^-$ generates a higher metastable HS state and ZFS in the [Fe(7*ditz*)$_3$](BF$_4$)$_2$ complex in contrast to the [Fe(6*ditz*)$_3$](BF$_4$)$_2$ case. An increase of the LIESST effect can be followed in the [Fe(8*ditz*)$_3$](BF$_4$)$_2$ complex, however without ZFS. Irradiation of the [Fe(9*ditz*)$_3$](BF$_4$)$_2$ compound during the LIESST experiment does not reach the same value as in the [Fe(9*ditz*)$_3$](ClO$_4$)$_2$ and the relaxation behaviour is slower.

3.6. PHOTO-MAGNETIC MEASUREMENTS

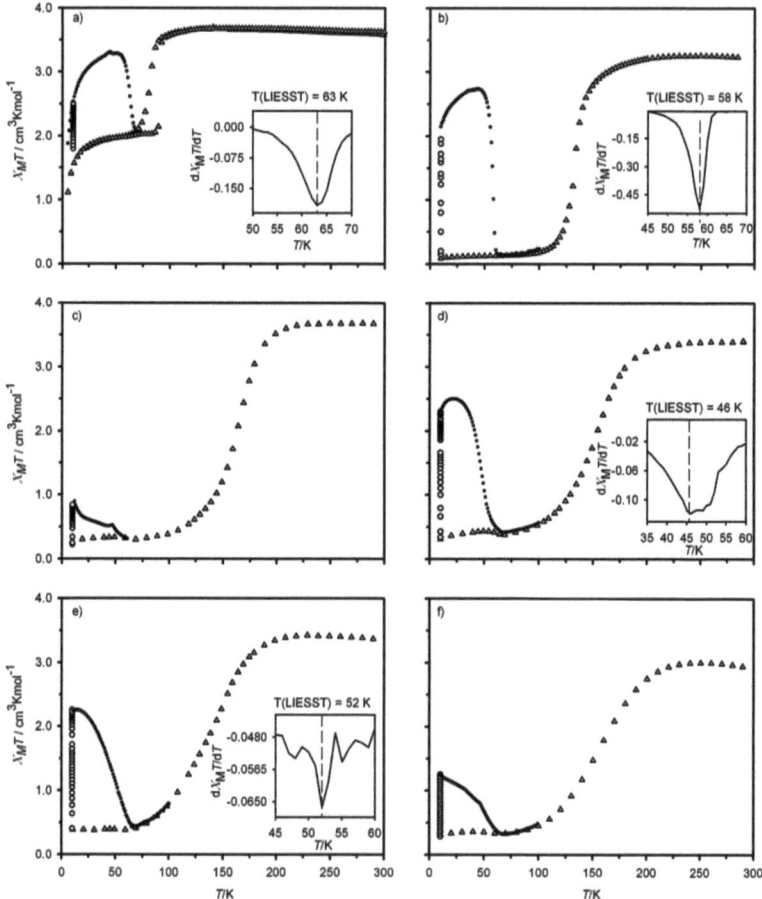

Figure 3.19.: Temperature dependence of $\chi_M T$ for (a) [Fe(4*ditz*)$_3$](BF$_4$)$_2$, (b) [Fe(5*ditz*)$_3$](BF$_4$)$_2$, (c) [Fe(6*ditz*)$_3$](BF$_4$)$_2$, (d) [Fe(7*ditz*)$_3$](BF$_4$)$_2$, (e) [Fe(8*ditz*)$_3$](BF$_4$)$_2$, (f) [Fe(9*ditz*)$_3$](BF$_4$)$_2$. (△) Data recorded without irradiation; (○) data recorded with irradiation at 10 K; (●) *T (LIESST)* measurement, data recorded in the warming mode with the laser turned off after irradiation for one hour. The existence of a small anomaly at around 50 K on *T(LIESST)* curve corresponds to an oxygen contamination even if particular precaution has been taken to purge the SQUID cavity for one hour at room temperature.

3.6. PHOTO-MAGNETIC MEASUREMENTS

[Fe(4$ditz$)$_3$](BF$_4$)$_2$

The lower $T_{1/2}$ of [Fe(4$ditz$)$_3$](BF$_4$)$_2$ results in a higher $T(LIESST)$ than [Fe(4$ditz$)$_3$](ClO$_4$)$_2$, although only 50 % of the switchable irons (LS state) are available. Furthermore, the light induced metastable HS state relaxes slower than that of [Fe(4$ditz$)$_3$](ClO$_4$)$_2$ and the ZFS is stronger pronounced. It is well known [82, 83, 84] that this light induced metastable HS state at low temperature can, in some cases, be populated through rapid freezing. [Fe(4$ditz$)$_3$](BF$_4$)$_2$ exhibits such an effect. If the sample is cooled down rapidly to 10 K from room temperature (taking approx. 5 min) a metastable HS fraction can be obtained. The temperature was then increased at a rate of 0.5 Kmin^{-1} and the metastable HS state starts to relax to the mixed HS-LS phase at 50 K before returning to the pure HS state at approximately 85 K.

If the initial cooling rate is decreased (0.5 Kmin^{-1}) then the transition occurs at 80 K and the mixed HS-LS state is stable down to very low temperatures (see Figures 3.17a and 3.18). Slow warming (1.0 Kmin^{-1}) follows now the same path back except the transition to HS occurs at 89 K equating to an hysteresis of approximately 10 K, which was also observed in the reflectivity experiments (see 3.3). Such behaviour has also been confirmed by Mössbauer experiments carried out at temperatures between 4.2 K and 294 K (see Figure 3.20 and Table 3.8). The isomer shifts are consistent for HS and LS iron(II) but have a wider quadrupole splitting than those previously measured for [Fe(4$ditz$)$_3$](PF$_6$)$_2$ [52].

3.6. PHOTO-MAGNETIC MEASUREMENTS

Table 3.8.: Isomer shift (δ) and Quadrupole Splitting (Δ) obtained for [Fe(4$ditz$)$_3$](BF$_4$)$_2$.

T/K	A_{LS} / %	δ_{LS} / mms^{-1}	Δ_{LS} / mms^{-1}	A_{LS} / %	δ_{LS} / mms^{-1}	Δ_{LS} / mms^{-1}
4.2	18.8	0.42	0.17	81.2	1.04	2.35
44	45.5	0.42	0.17	54.5	1.05	2.43
70	44.0	0.41	0.19	56.0	1.05	2.47
110	6.2	0.32	0.44	93.9	1.02	2.34
294	5.80	0.30	0.58	94.2	0.92	1.64

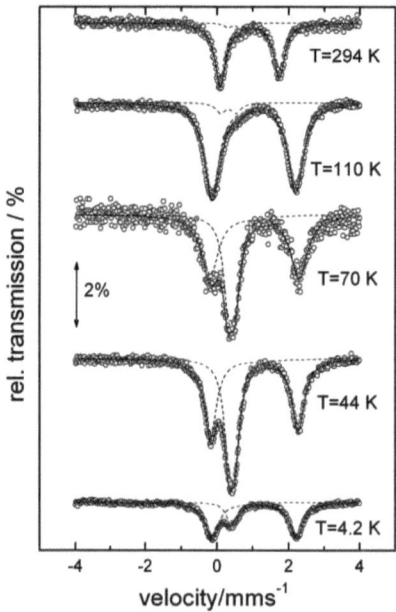

Figure 3.20.: ^{57}Fe-Mössbauer spectra of [Fe(4$ditz$)$_3$](BF$_4$)$_2$ at selected temperatures.

3.7. Low-spin/High-spin Transition in Megagauss Fields

To investigate the influence of high magnetic fields in a SC complex that shows hysteresis and a sharp two step transition, the [Fe(4*ditz*)$_3$](PF$_6$)$_2$.EtOH complex has been investigated. As already described in section 3.2 the material exhibit a pronounced visible color change in the temperature range between 77 K and 300 K indicating the high-spin/low-spin phase transition. The principal objective of the experiments was the demonstration that this phase transition and hence change in color can also be obtained by application of strong magnetic fields in the megagauss range, since lower magnetic fields up to 40 T proved to be not successful for the discussed material. The sample was checked for its exact temperature dependence of the phase transition to find the optimal temperature for the experiment in megagauss fields. In Figure 3.21 both the reflection data and the magnetic field intensity as a function of time in the upper and lower part respectively for [Fe(4*ditz*)$_3$](PF$_6$)$_2$.EtOH for λ = 632 nm at T = 183.4 K are plotted. The second wavelength λ = 541 nm provided no additional information. The experimental data of the reflection (upper solid curve in Figure 3.21) follows the experimental magnetic field (lower solid curve) with a pronounced hysteresis. This means, that the relaxation times involved are of the order of μ sec. There is a clear increase of the reflectivity with increasing magnetic field indicating the increased population of the high-spin levels. The broken curves indicate the results of the corresponding simulation discussed below. To demonstrate both the magnetic field dependence and the hysteresis involved the data were plotted directly as a function of the applied magnetic field (Figure 3.22). Again solid and broken curves indicate experimental and simulation results, respectively.

3.7. SPIN CROSSOVER IN MEGAGAUSS FIELDS

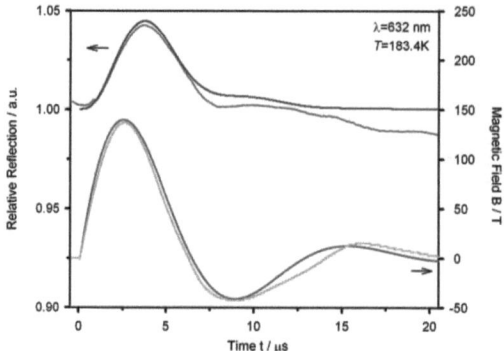

Figure 3.21.: Experimental data (– and –) and simulation (– and –) for the relative reflection and the magnetic field as a function of time.

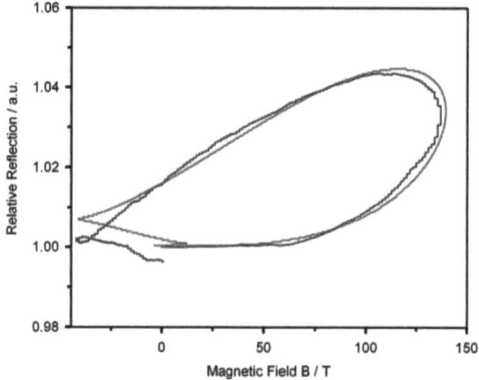

Figure 3.22.: Experimental data (–) and simulation of the hysteresis.

3.7. SPIN CROSSOVER IN MEGAGAUSS FIELDS

3.7.1. Variation of the Temperature at Constant Field

The [Fe(4$ditz$)$_3$](PF$_6$)$_2$.EtOH complex was measured at different temperatures and an applied field of 140 T (coil dimension:12mm×12mm×3mm, charging voltage: 40 kV). It was possible to induce the spin transition at 172.6, 183.4 and 186.5 K (see Figure 3.23). The applied field stabilises the induced HS state and hence with rising temperature the population of the HS states also increases, not only because of the applied field but also because of the thermally induced SC. At 177 and 165 K a magnetic induced SC was not detectable.

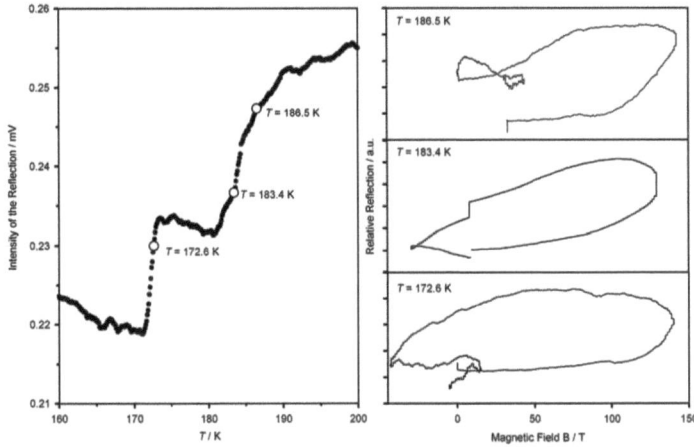

Figure 3.23.: Effect of the induced magnetic field on [Fe(4$ditz$)$_3$](PF$_6$)$_2$.EtOH complex at different temperatures.

3.7. SPIN CROSSOVER IN MEGAGAUSS FIELDS

3.7.2. Variation of the Magnetic Field

To show the influence of variable high magnetic fields ($\frac{dB}{dT}$) on the SC behaviour of [Fe(4$ditz$)$_3$](PF$_6$)$_2$.EtOH at 183.4 K, the following coil dimensions were used: (i) 12mm × 12mm × 3mm (ii) 13mm × 13mm × 3mm (iii)15mm × 15mm × 3mm (see Figures 3.24-3.26). The applied magnetic fields are in the range of 100 to 160 T. This series of measurements show that an applied field of 140 T caused the highest population of the HS state at the selected temperature. At higher fields, the excitation decreases again. The slew rate of the magnetic field is directly linked to the spin transition.

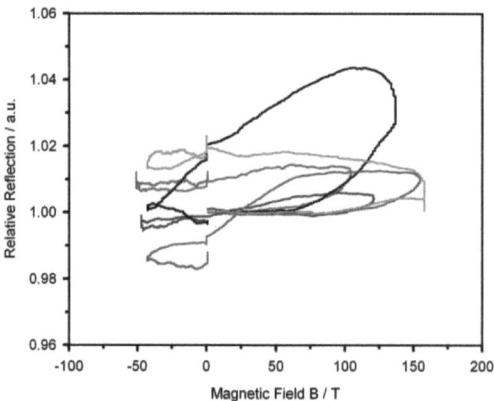

Figure 3.24.: Excitation of the SC in [Fe(4$ditz$)$_3$](PF$_6$)$_2$.EtOH at 183.4 K at different induced magnetic fields with the coil of 12mm× 12mm× 3mm and different charging voltages: (–) 30 kV,(–) 35 kV,(–) 40 kV,(–) 45kV and (–) 50 kV.

3.7. SPIN CROSSOVER IN MEGAGAUSS FIELDS

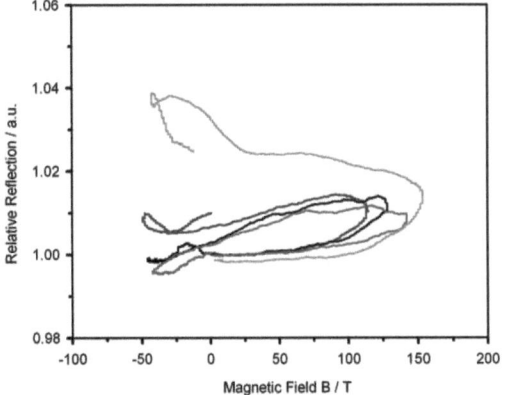

Figure 3.25.: Excitation of the SC in [Fe(4$ditz$)$_3$](PF$_6$)$_2$.EtOH at 183.4 K at different induced magnetic fields with the coil of 13mm× 13mm× 3mm and different charging voltages: (–) 35 kV,(–) 40 kV,(–) 45 kV and (–) 50 kV.

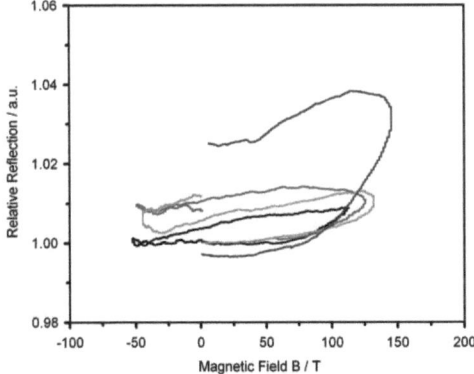

Figure 3.26.: Excitation of the SC in [Fe(4$ditz$)$_3$](PF$_6$)$_2$.EtOH at 183.4 K at different induced magnetic fields with the coil of 15mm× 15mm× 3mm and different charging voltages: (–) 40 kV,(–) 45 kV,(–) 50 kV and (–) 50 kV.

Discussion

4. Discussion

4.1. Powder Diffraction

From the visual appearance of the powder patterns the [Fe(n*ditz*)$_3$](BF$_4$)$_2$ complexes split into two groups, namely the phase with the butylene spacer and the series with n = 5-10 and 12. This division also holds for the corresponding perchlorate compounds [Fe(n*ditz*)$_3$](ClO$_4$)$_2$ whereupon these show a higher degree of cristallinity compared to the tetrafluoroborates. Comparative inspection of the tentative intensities of the [Fe(4*ditz*)$_3$](BF$_4$)$_2$ and [Fe(4*ditz*)$_3$](PF$_6$)$_2$ [52] complex, clarifies that the BF$_4^-$ complex constitutes a distorted variant of the PF$_6^-$ salt. Unfortunately the present powder quality does not allow structure solution from the pattern and is subject to currently ongoing research. The interpretation of the [Fe(4*ditz*)$_3$](ClO$_4$)$_2$.EtOH phase's pattern is less clear-cut but it can be said that it is also structurally closely related to the BF$_4^-$ and PF$_6^-$ complexes. Due to the high reflection density, however, indexing of such a complex pattern is outside the scope of laboratory X-ray powder data. The apparent difference in the unit cells between the presented product and the phase reported earlier [58] is ascribed to the different solvents used to crystallise the complexes. Whereas van Koningsbruggen *et al.* [58] used methanol the present samples were precipitated from ethanol. Concerning the homologous series [Fe(n*ditz*)$_3$](X)$_2$ with X = BF$_4^-$, ClO$_4^-$ and n = 5-10 and 12 it was possible to index the powder pattern of the [Fe(5*ditz*)$_3$](BF$_4$)$_2$ complex. The obtained parameters can be regarded as a distorted variant of the trigonal symmetry found for [Fe(2*ditz*)$_3$](BF$_4$)$_2$ [56] with comparable lattice parameters. Therefore it is obvious to describe the pentylene- to dodecylene-ditetrazole complex series in a chain-type arrangement as well. The higher order complexes are suspected

4.1. POWDER DIFFRACTION

to crystallise in the triclinic system which, however, is not possible to confirm from the low quality powder patterns. Calculating the unit cell volumes for the phases under investigation, assuming the monoclinic cell from above and setting the b-axis length to d × 2 of the longest line (i.e. the 020 reflection), results in unit cell contents Z of about 2, which again is in accordance with the $[Fe(2ditz)_3](BF_4)_2$ complex. In Figure 4.1 an idealised tentative structural proposition is depicted. The ligands, linked together by iron(II), are stretched out and aligned parallel in a hexagonal close-packed manner with the counter ions in between the chains. Increase of the alkylene spacer length will expand the unit cell in the crystallographic direction of the chains leaving the other two untouched. Additionally the voids between them become larger and are likely to be filled with presumably disordered and non-stoichiometric amounts of solvent molecules. As the interaction between the chains is weak (the Coloumb forces between the positively charged iron(II) and BF_4^- will diminish with increasing n) stacking faults are favoured by longer spacers and additional reflection broadening is likely to be observed.

4.1. POWDER DIFFRACTION

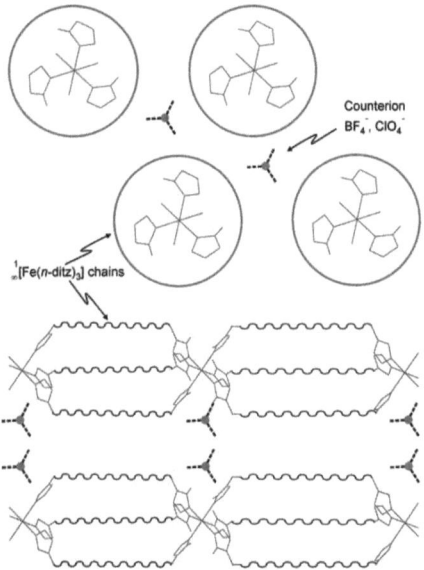

Figure 4.1.: Tentative basic structural model of the [Fe(n*ditz*)$_3$](X)$_2$.EtOH (n = 5-10, 12, X = BF$_4^-$ and ClO$_4^-$) phases viewed along the [Fe(n*ditz*)$_3$]$^{2+}$ chains (circled, top) and the perpendicular direction (bottom). The orientation and geometry of the complex is presumed arbitrarily.

4.2. UV/VIS-NIR Spectroscopy

The members of the discussed compounds, namely the [Fe(n$ditz$)$_3$](BF$_4$)$_2$, [Fe(n$ditz$)$_3$](ClO$_4$)$_2$ with n = 4-9 and the [Fe(4$ditz$)$_3$](X)$_2$ with X = BF$_4^-$, ClO$_4^-$, PF$_6^-$, SbF$_6^-$ and ReO$_4^-$, show a thermochromic effect associated with the spin transition, from white in the HS state and violet in the LS state. The spin transition is detected in terms of three peaks, two of which decrease and one that increase with increasing temperature. These peaks can be assigned to the spin allowed transitions: $^1A_1 \rightarrow {}^1T_2$, $^1A_1 \rightarrow {}^1T_1$ and $^5T_2 \rightarrow {}^5E$ whereas these d-d transitions can be found found at $\approx$ 26000 cm^{-1}, 18200 cm^{-1} and 12000 cm^{-1}. Due to the cooling system of the spectrophotometer, the displayed temperatures are shifted by a value of 15 K in comparison to the magnetic measurements. Therefore the emphasis is given on the analysis of the obtained absorption bands and the ligand field properties of the measured compounds.

Regarding the HS band the discussion must be focused on the appearance and the parameters obtained in this investigations. The asymmetric appearance of the peak is the most conspicuous property. This asymmetry was already seen in the related compounds [Fe(mtz)$_6$](BF$_4$)$_2$,[Fe(etz)$_6$](BF$_4$)$_2$ and [Fe(ptz)$_6$](BF$_4$)$_2$ (mtz = 1-methyl-1H-tetrazole, etz = 1-ethyl-1H-tetrazole, ptz = 1-propyl-1H-tetrazole) [15, 48, 49, 85]. The anomaly of the HS band in the complexes [Fe(mtz)$_6$](BF$_4$)$_2$ and [Fe(etz)$_6$](BF$_4$)$_2$ is explained by the structurally properties of the compounds. The two compounds possess two different iron(II) lattice sites A and B, whereupon one iron(II) stays in the HS state until 4.2 K and the other iron(II) exhibits a spin transition to LS under 160 K. These results suggest that the unusual form of the HS band in the absorption spectra of these two compounds can be attributed to the two nonequivalent iron(II) lattice sites.

4.2. UV/VIS-NIR SPECTROSCOPY

By virtue of the structures of the propyltetrazole complex, the [Fe(4$ditz$)$_3$](PF$_6$)$_2$ complex as well as for the [Fe(4$ditz$)$_3$](BF$_4$)$_2$ and [Fe(4$ditz$)$_3$](ClO$_4$)$_2$ complex the reason for the appearance of the HS band of the discussed complexes in this work cannot be explained with two different iron(II) latice sites. As already mentioned by Decurtins et al. [48] an alternative explanation for this unusual property is the Jahn-Teller distortion. In 1937 Jahn and Teller showed that in general no nonlinear molecule can be stable in a degenerated electronic state. The molecule must become distorted in such way as to break down the degeneracy. Such distortions may be evident from the spectroscopic properties of the complex [86, 87]. The theorem holds for the ground states as well as for the excited states, whereas the explanation in the latter case is more difficult because the short lifetime of the excited state avoids the adjustment of a stable equilibrium. For example, the electron configuration of [Fe(H$_2$O)$_6$]$^{2+}$ is t$_{2g}^4$e$_g^2$ for the ground state and t$_{2g}^3$e$_g^3$ for the excited state, with the same number of unpaired electrons. Therefore the excited state of this ion is Jahn-Teller distorted and consequently split into two species. This split can then be detected in the absorption spectra. In Figure 4.2 the absorption spectra of [CoF$_6$]$^{-3}$ in K$_2$Na[CoF$_6$] is taken as an example for such a result. Interestingly, these Jahn-Teller distortions can also be found for compounds with 4 electrons in the t$_{2g}$ orbital hence in the HS state of the discussed series.

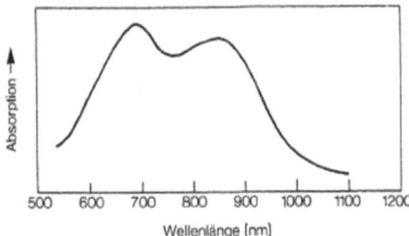

Figure 4.2.: Absorption spectra of the [CoF$_6$]$^{-3}$ in K$_2$Na[CoF$_6$], the splitting of the t$_{2g}^3$e$_g^3$ excited state as a result of the Jahn-Teller distortion is visible [86]

4.2. UV/VIS-NIR SPECTROSCOPY

The broad HS band seen in the absorption spectra of the complexes can now be considered as an overlap of two bands, which unfortunately due to the powder form and measurement method of the sample cannot be distinguished or deconvoluted. Consequently the HS band was fitted with one Gaussian curve, described in chapter 3.2.1 and equation 3.2. From the above mentioned results for the Jahn-Teller distortion of Fe(II) the width of the fitted curve must be directly correlated to the dimension of the distortion. Because of the fact that obviously these two bands do not disappear synchronously, the band appears to decrease at first only on the right side. But if one considers two bands, where one decreases faster than the other, the shift of the maximum and the asymmetry is understandable. If the obtained values for the HS band widths of the two series [Fe(n*ditz*)$_3$](BF$_4$)$_2$ and [Fe(n*ditz*)$_3$](ClO$_4$)$_2$ are compared (note, due to the structural difference of the [Fe(4*ditz*)$_3$](X)$_2$ with X = BF$_4^-$ and ClO$_4^-$ complexes these are excluded from this comparison), an alternating dependence between n and the width is visible in the ClO$_4^-$ complexes (see Figure 4.3 a,b).

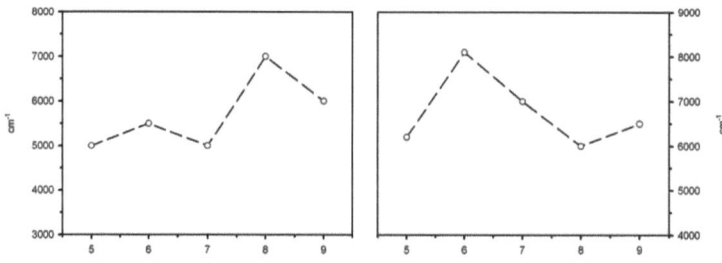

Figure 4.3.: Dependence between n and the width of the HS band for a) [Fe(n*ditz*)$_3$](ClO$_4$)$_2$ with n = 5-9 and b) [Fe(n*ditz*)$_3$](BF$_4$)$_2$ with n = 5-9 complexes.

This means that in the ClO$_4^-$ series, the complexes with the even number of carbons in the spacer are more distorted than the ones with the odd number of carbons, whereas [Fe(8*ditz*)$_3$](ClO$_4$)$_2$ shows the highest degree of distortion. The BF$_4^-$ series show a similar trend whereupon the [Fe(8*ditz*)$_3$](BF$_4$)$_2$ complex exhibits again the exception: the parity is broken and it is the complex with

4.2. UV/VIS-NIR SPECTROSCOPY

the lowest degree of distortion. This obtained parity effect of the two series is one of the fascinating results, concerning the investigated compounds. The same analysis is made for the [Fe(4*ditz*)$_3$](X)$_2$ complexes with X = BF_4^-, ClO_4^-, PF_6^-, SbF_6^- and ReO_4^-. If one looks at the obtained values for the width of the HS band at RT (see Table 3.4) the ClO_4^- and ReO_4^- complexes reach much higher values compared to the BF_4^-, PF_6^- and SbF_6^- which have the same values. Therefore it can be concluded that the complexes are divided into two groups the ones with a fluorin containing anion and the ones with an oxygen containing anion. This could lead to the assumption that the ClO_4^- and the ReO_4^- anion leads to higher Jahn-Teller distortion than BF_4^-, PF_6^- and SbF_6^-. The calculated ligand field strength Δ_O of the discussed BF_4^- and ClO_4^- series as well as for the [Fe(4*ditz*)$_3$](X)$_2$ complexes are in the expected region compared to the related monotetrazole compounds [15, 48, 49, 85]. For neutral ligands such as the n*ditz*, Δ_O is expected to vary as $\approx 1/r^6$ [88, 89], where r is the metal-ligand distance. Therefore the proportionality $\Delta_O(LS) / \Delta_O(HS) \approx (r_{HS} / r_{LS})^6$ should hold, with r_{HS} and r_{LS} being the Fe-N bondlengths in the HS and LS state, respectively. As the the crystal structure of the [Fe(4*ditz*)$_3$](PF$_6$)$_2$ and [Fe(4*ditz*)$_3$](ClO$_4$)$_2$ are solved, these two complexes were used to verify this proportionality (see equation 4.1 and 4.2). (Note that, due to the very low transition temperature of the [Fe(4*ditz*)$_3$](BF$_4$)$_2$ complex, the energy of the LS band cannot be determined. Therefore, even though, the crystal structure of the complex is solved, it can not be taken for the calculation.) r_{LS} and r_{HS} of the [Fe(4*ditz*)$_3$](ClO$_4$)$_2$ complex are 2.002 Å and 2.191 Å[1] and the calculated ligand field strengths for LS and HS are 20056 and 11500 cm^{-1}.

$$\frac{\Delta_0^{LS}}{\Delta_0^{HS}} = 1.74 \qquad (4.1)$$

$$(\frac{r_{HS}}{r_{LS}})^6 = 1.72 \qquad (4.2)$$

The values for [Fe(4*ditz*)$_3$](PF$_6$)$_2$ are: r_{HS} and r_{LS} 2.193 Å and 2.00 Å and the

---
[1] Due to the fact that in the LS and HS state of the [Fe(4*ditz*)$_3$](ClO$_4$)$_2$ complex different Fe-N bondlengths are detected [62] an average of these values has been taken.

ligand field strengths for LS and HS are 20056 and 11700 cm^{-1}. The ratio of the ligand field strength is therefore 1.71 and 1.73 is the result of $(r_{HS}/r_{LS})^6$. This apparently perfect agreement should not be overrated, but it is a confidence, however, ligand field theory can be used in a first step for describing the absorption spectra.

4.3. Magnetic and Photo-Magnetic Measurements

In the following section the results of the magnetic and photo-magnetic measurements of the [Fe(n$ditz$)$_3$](BF$_4$)$_2$ and [Fe(n$ditz$)$_3$](ClO$_4$)$_2$ series will be discussed and compared with other performed measurements. The aim of this discussion is to clarify the influence of the spacer length of the ligand as well as the influence of the anion on the obtained properties of the series. To simplify matters the structurally different [Fe(4$ditz$)$_3$](X)$_2$ complexes with X = BF$_4^-$ and ClO$_4^-$ will be discussed separately after the disquisition on the [Fe(n$ditz$)$_3$](BF$_4$)$_2$ and [Fe(n$ditz$)$_3$](ClO$_4$)$_2$ complexes.

[Fe(n$ditz$)$_3$](ClO$_4$)$_2$ Series

Firstly I will focuse on the temperature of the thermal spin-state transition of the [Fe(n$ditz$)$_3$](ClO$_4$)$_2$ series with n = 5-9, $T_{1/2}$, estimated from the maximum in the first derivative of the $\chi_M T$ versus T plot which are given in Table 3.7, and Figure 4.4 shows the change of $T_{1/2}$ with the length of the alkylene spacer (n). Interestingly, an increase in the length of the spacer raises the $T_{1/2}$ value, but it can also be seen that the influence of the parity of the spacer is not negligible. Indeed, the complexes can be divided into two series depending on the parity of the spacer (see Figure 3.15): complexes with odd-numbered carbon atoms in the spacer, reach a lower $\chi_M T$ product at room temperature than the ones with even numbered ligands.

4.3. MAGNETIC AND PHOTO-MAGNETIC MEASUREMENTS

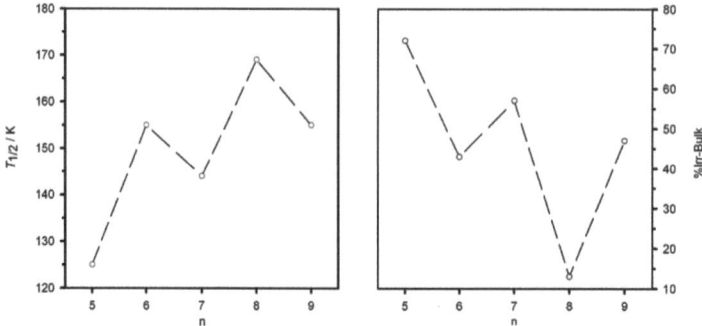

Figure 4.4.: $T_{1/2}$(a) and %Irr-Bulk (b) as a function of the number of carbons n in the spacer of the ditetrazole ligands for the perchlorate series.

To verify this separation of the complexes depending on the number of carbons n in the spacer, the calculated amount of photo-bleached fraction (%Irr-Surface; reported in Table 3.7) for each compound relative to the value found at room temperature, is also correlated to n, whereas this comparison shows that the level of photo-excitation roughly decreases with the length of the spacer. It cannot be excluded that other parameters, such as the morphology of the powder particles, should be taken into account in order to discuss the problem of the light penetration, but it is clear that the level of photo-excitation follows the increase of the $T_{1/2}$ value: the higher the $T_{1/2}$, the lower is the level of the light-induced LS/HS conversion. This result can, in fact, be discussed in terms of the inverse energy-gap law introduced by Hauser [38]. In the so-called single configurational coordinate (SCC) model, the potential wells of the LS and HS states are plotted along a single reaction coordinate Q, which describes the totally symmetric breathing mode and is related to the metal–ligand bond difference by $\Delta Q = \sqrt{6}\Delta r$ (see Figure 4.5). The potential wells and their shape can, therefore, be moved horizontally relative to each other by changing the relative bond strengths of HS and LS, or the wells can be moved vertically relative to each other. The difference in the zero-point energies of LS and HS, ΔE, is therefore related to $T_{1/2}$. The recorded far-FTIR spectra of all the members of the

4.3. MAGNETIC AND PHOTO-MAGNETIC MEASUREMENTS

[Fe(n*ditz*)$_3$](ClO$_4$)$_2$ series show almost no difference (see Table 3.6). Strikingly, the LS state shows a band that only varies between 421 and 424 cm^{-1} and the HS state shows bands that vary by a maximum of 10 cm^{-1} and certainly do not follow any obvious trend. From that it can be concluded that the shape of the potential wells of the LS and the HS states, as well as the horizontal distance between them (ΔQ), can be considered almost the same for all complexes. In other words, along the [Fe(n*ditz*)$_3$](ClO$_4$)$_2$ series the change in $T_{1/2}$ implies a vertical displacement of the two potential wells and hence a variation in ΔE with the parity, as shown in Figure 4.4. A further consequence of the variation of ΔE can be explained as it has previously been shown that there is a direct relationship between the rate constant of tunnelling, k_o, and $T_{1/2}$ [90]. Based on this finding one can see that if the stability of the photo-induced HS state decreases with the increase of $T_{1/2}$ (and of ΔE), for a constant intensity of light irradiation, the photo-excitation becomes more and more difficult, as observed experimentally (%Irr-Surface; reported in Table 3.7).

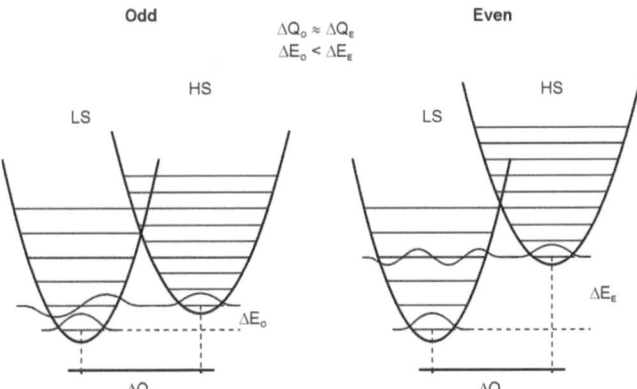

Figure 4.5.: Potential wells for an odd and even ditetrazole coordination polymer demonstrating the effect of the parity of the bridging ligand. Only ΔE appears to be affected and not ΔQ.

4.3. MAGNETIC AND PHOTO-MAGNETIC MEASUREMENTS

The photo-magnetic properties of the different compounds were investigated by following the effect of irradiation with light under the influence of an applied magnetic field. Due the fact that one procedure was followed for all investigated compounds, the quantification of $T(LIESST)$, which is determined by the minimum of the $\delta\chi_M T/\delta T$ vs. T curve recorded during relaxation becomes possible. Table 3.7 collects, for each compound, the highest percentage of photo-conversion obtained by irradiation at 10 K relative to the magnetic value recorded at room temperature for a pure HS state (%Irr-Bulk). As expected from the surface analysis, one can see here that with bulk detection the level of photo-excitation is not uniform for all the compounds, even when special attention was paid to tune the intensity and/or the wavelength. This confirms once more that the level of photo-excitation decreases with an increase of $T_{1/2}$ and therefore clearly shows the influence of both the length and parity of the bridging ligand (see Figure 4.4). This behaviour provides new evidence that the stability of the photo-induced HS state in the tunnelling region varies along the $[Fe(nditz)_3](ClO_4)_2$ series. In fact, only for complexes possessing a sufficiently long lifetime does the photo-induced HS fraction remain almost independent of any change of the temperature and of the effect of time during the measurement of the $T(LIESST)$ curve.

4.3. MAGNETIC AND PHOTO-MAGNETIC MEASUREMENTS

[Fe(n*ditz*)$_3$](BF$_4$)$_2$ Series

To verify this fascinating result of the parity found in the ClO$_4^-$ series, a similar analysis was completed for the BF$_4^-$ complexes. As in the ClO$_4^-$ complexes also for the BF$_4^-$ complexes the amount of photo-bleached fraction from the reflectivity measurements (%Irr-Surface; reported in Table 3.7) was calculated. However, similar to the previously discussed ClO$_4^-$ analogues, the higher the $T_{1/2}$, the lower the level of the light-induced LS/HS transformation on the surface. This relation can also be found in the comparison of the magnetic and magneto-optical data, presented in Table 3.7. Comparison of the new magnetic data with the perchlorate data shows a similar trend from n = 5-7. The temperatures of transition, $T_{1/2}$, are all higher for BF$_4^-$ than for ClO$_4^-$. The %Irr-Surface and %Irr-Bulk values show a similar relationship with n but the actual values are different suggesting a similar relationship with parity. There is, however, a disjunction when n is changed from 7 to 8 (see Figure 4.6). This reversal of the parity dependence at this point suggests the smaller BF$_4^-$ causes a change in the arrangement of the alkyl chains. Rising n up to 12 the transitions become remarkably similar. This might be because any further changes in the arrangement of the alkyl chains would now be insignificant with respect to the greater separation of the iron centres. This suggests that as the alkylene spacer is lengthened $T_{1/2}$ eventually reaches a limiting value of 160 K. As mentioned in 3.1, perchlorate samples in general crystallize more readily than the tetrafluoroborate analogues. It is, however, obvious that the tetrafluoroborates show a limited parity effect from n = 5-7, however if n is increased from 7 to 8 the parity effect is reversed (see Figure 4.6). This suggests that at this length the alkylene chains are differently arranged.

This result of the parity is repeated, if the width of the $^5T_2 \rightarrow {^5E}$ transition is displayed as function of n (see Figure 4.3). Actually even the reverse of the alternating parity at [Fe(8*ditz*)$_3$](BF$_4$)$_2$ appears again. This agreement of the results, lead to the conclusion that a small change in the spacer length as well as in the size of the anion has an remarkable influence on the properties of the presented complexes.

4.3. MAGNETIC AND PHOTO-MAGNETIC MEASUREMENTS

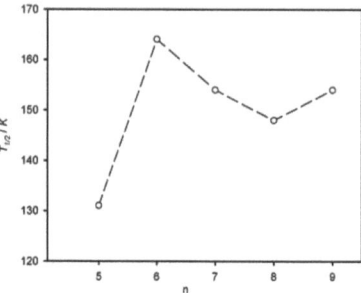

Figure 4.6.: Comparison of $T_{1/2}$ as function of the number of carbons (n) in the spacer of the ditetrazole ligands for the BF_4^- series.

[Fe(4$ditz$)$_3$](BF$_4$)$_2$ and [Fe(4$ditz$)$_3$](ClO$_4$)$_2$

As already mentioned during the previous discussions, the complexes with the butylene spacer as bridging ligand behave, in respect of the mentioned properties, somehow differently in comparison to the complexes with the longer spacers. Regarding the thermal spin transition [Fe(4$ditz$)$_3$](ClO$_4$)$_2$ is the only one within the ClO$_4^-$ series that displays a two-step thermal spin-transition with a plateau between 93 and 124 K. This behaviour can be compared with the same complex synthesised in methanol solution and published previously by van Koningsbruggen et al. [58]. The magnetic susceptibility curves of the two [Fe(4$ditz$)$_3$](ClO$_4$)$_2$ compounds (one synthesised in methanol and the other synthesised in ethanol) show clear differences in both the shape of the spin-transition curves and in their spin-transition behaviour. Indeed, the shape of the magnetic susceptibility curve of the compound made previously in methanol shows an incomplete one-step spin transition. This suggests a significant influence of the solvent on this coordination polymer (see [62] and 3.2.3). An extensive discussion of this effect can be found in [62]. It is also important to note that the peculiarity of the [Fe(4$ditz$)$_3$](ClO$_4$)$_2$ spin-transition, shows similar features in the T(LIESST) experiments. Analysis of the $\delta\chi_M T/\delta T$ vs. T curve clearly shows the existence of two minima, at 39 K and 58 K, thus proving that both magnetically non-equivalent iron(II) metal centres can be

4.3. MAGNETIC AND PHOTO-MAGNETIC MEASUREMENTS

photo-excited. The presence of two minima on the $T(LIESST)$ curve is not unexpected as a similar result has been found for [Fe(DPEA)(bim)](ClO$_4$)$_2$·0.5H$_2$O (DPEA = (2-aminoethyl)bis(2-pyridylmethyl)amine, bim = 2,2'-bisimidazole) which also displays a two-step thermal spin-transition [91].

The [Fe(4$ditz$)$_3$](BF$_4$)$_2$ complex exhibits in comparison to the [Fe(4$ditz$)$_3$](ClO$_4$)$_2$ complex different properties, which can already be detected in the reflectivity measurement. There it shows a strong pronounced LITH and coexistently a shift of the reflection maximum by 10 K. This seems to indicate the presence of a thermal hysteresis during the spin transition which is confirmed by the SQUID measurements described. The existence of an hysteresis associated to the LIESST phenomenon corresponds to the LITH defined by Létard *et al.* [92] and correlated by Varret *et al.* [41] to the existence of cooperative interaction in the lattice. Such a result is confirmed by the existence of an hysteresis loop on the thermal SC regime. The [Fe(4$ditz$)$_3$](BF$_4$)$_2$ complex shows an incomplete SC of approximately 50 % of the iron centres with an hysteresis of 9 K and a second sharp decrease of $\chi_M T$ below 25 k is observed. This drop might be associated with a second spin transition at around 5 K. This would mean a shift of both SC temperatures by some 50 - 60 K. However this sharp decrease can also be associated with ZFS. The latter explanation seems to be more likely as the spin state should be frozen at such low temperatures. A similar phenomenon has also been observed by Klingele *et al.* [93] using a 1,2,4-triazole-bridged Fe(II) complex with BF$_4^-$ as its anion and Real *et al.* [94, 95] when discussing the dinuclear [Fe(bpym)(NCSe)$_2$]$_2$ compound. The lower $T_{1/2}$ of [Fe(4$ditz$)$_3$](BF$_4$)$_2$ results in a higher $T(LIESST)$ than [Fe(4$ditz$)$_3$](ClO$_4$)$_2$, although, of course, only 50 % of the switchable irons (LS state) are available. Furthermore, the light induced metastable HS state relaxes slower than that of [Fe(4$ditz$)$_3$](ClO$_4$)$_2$ and the ZFS is stronger pronounced. It is well known [82, 83, 84] that this light induced metastable HS state at low temperature can, in some cases, be populated through rapid freezing. [Fe(4$ditz$)$_3$](BF$_4$)$_2$ exhibits such an effect which is not only proved in the SQUID measurement but also with Mössbauer. To compare these results with the SQUID data the latter was converted into mole

4.3. MAGNETIC AND PHOTO-MAGNETIC MEASUREMENTS

fraction (see equation 4.3) and plotted in Figure 4.7.

$$mole fraction = \frac{\chi_M T - (\chi_M T)_{min}}{(\chi_M T)_{max} - (\chi_M T)_{min}} \qquad (4.3)$$

The mole fraction values calculated from the Mössbauer data are within experimental error the same as the SQUID data obtained after fast cooling except for one point measured at 44 K. In preparation for the Mössbauer experiments the sample is cooled down to 4.2 K within 20 minutes for the first experiment and then the sample is slowly heated for each subsequent experiment. The measurement time at each temperature averages 8 days. The spectrum measured at 4.2 K shows that cooling of the sample is fast enough that most of the sample remains in the HS state (approx. 80%). It also shows that at this temperature the relaxation time of this metastable HS state has a $t_{1/2} > 8$ days and this is confirmed by analysis of Mössbauer spectra recorded during the 8 days which add together to give the final spectrum given in Figure 3.20. At 44 K, however, the Mössbauer data shows that during the long measurement time the system has relaxed back to the mixed LS-HS state: indeed spectra taken every 12 hours show that $t_{1/2} \approx 2$ days. Complete relaxation during Mössbauer experiments even at 4.2 K has been observed by Yamada et al. [82] when investigating a Fe(II) complex based on a tripodal ligand with coordinating imidazoles ([Fe^{II}H$_3$L^{Me}]Cl.I$_3$ - (H$_3$L^{Me} = tris[2-(((2-methylimidazoyl-4-yl)methylidene)amino)ethyl]amine)). This complex crystallises in a 2D layer structure linked by NH-Cl- hydrogen bonds between the Cl$^-$ ion and three neighbouring imidazole groups. The magnetic measurement shows a metastable HS state caused by rapid cooling but the Mössbauer measurements do not. It seems that the weak inter-layer linking caused by the Cl$^-$ ion means the relaxation time from the metastable HS state is faster than the Mössbauer measurement time. In related compounds Matsumoto and co-workers [83, 84] show 2D networks consisting of two tripodal components tied together by imidazole-imidazolate hydrogen bonds. Here the metastable HS state is observable in both SQUID and Mössbauer measurements. In these cases the relaxation time is longer than the measurement times and this could

4.3. MAGNETIC AND PHOTO-MAGNETIC MEASUREMENTS

be because of the stronger hydrogen-bond interactions within the layer.

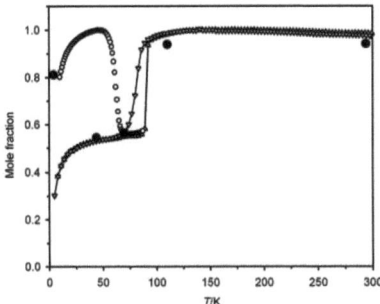

Figure 4.7.: Comparison of the mole fraction of [Fe(4*ditz*)$_3$](BF$_4$)$_2$ from SQUID (slow cooling ▽; slow warming △; warming preceded by rapid cooling ○) and Mössbauer measurement warming preceded by rapid cooling (●).

4.4. High Magnetic Field Measurements

Generally the SC can also be induced by the different magnetic-field dependence of the spin levels involved, so that beyond a critical magnetic field the population of the spin-levels is inverted. In this way temperature, and magnetic field are complementary parameters to control the system with respect to the majority population and its related effects experimentally.The principal objective of the presented experiments was the demonstration that this phase transition and hence change in color can also be obtained by application of strong magnetic fields in the megagauss range. To detect the HS/LS transition the applied setup is measuring the normal reflection in Faraday configuration using Plastic Optical Fiber (*POF*) for the monochromatic laser radiation of λ = 632 nm. The corresponding sample holder was mounted in a miniature N_2-cryostat to meet the limited dimensions of the magnetic field coil. As detector a fast photo-diode with 125 MHz bandwidth was used. As magnetic field generator a single-turn coil was used, providing in the experiments peak fields of 160 T for a half-sine pulse of the order of 6 μsec length. The single-turn coils of 12 mm and 15 mm diameter were driven by a 225 kJ/60 kV capacitor bank providing a peak current of the order 3 MA [67]. The magnetic field was measured by a calibrated pick-up coil with suitable integrator. Both data channels were set at 100 MHz sampling frequency.

4.4.1. Simulation

In general the calculations for the induction of the SC through a high magnetic field are made for fields smaller than 100 T. The usually simplification assumed $g\mu_B B \ll k_B T$ can not be used for such high induced fields used in the described experiments. Due to the two possible spin states S = 0 for the LS state and respectively S=2 for the HS state the calculation is done for a two split level system. For B = 0 the HS state is fivefold degenerated whereas the LS state is not degenerated (see Figure 4.8).

4.4. HIGH MAGNETIC FIELD MEASUREMENTS

Based on the electron dispersion in the e_g and t_{2g} orbitals the equilibrium energies for Δ_O LS and Δ_O HS are calculated as follows:

$$\Delta_0(LS) = -\frac{6}{15} 10Dq(LS) \quad (4.4)$$

$$\Delta_0(HS) = -\frac{1}{15} 10Dq(HS) \quad (4.5)$$

The energy gap between the S=0 and S=2 state is the difference between the equilibrium energies:

$$\Delta_0 = \Delta_0(HS) - \Delta_0(LS) \quad (4.6)$$

In an magetic field, the energy-levels of the HS state split because of the Zeeman-Effect and the distance between the spin-levels changes (see Figure 4.8).

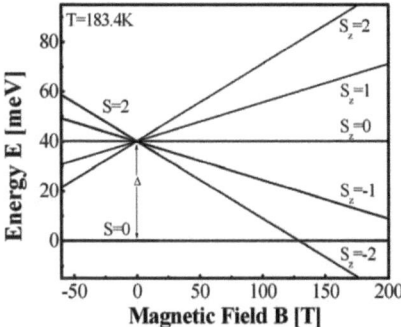

Figure 4.8.: Energy-level system as a function of the magnetic field.

4.4. HIGH MAGNETIC FIELD MEASUREMENTS

To simulate the experimental results the model of this split (S=2/S=0) level system as shown in Figure 3.26 which corresponds to the Hamiltonian [96] (see equation 4.7) is applied.

$$H = (\Delta + g\mu_B S_z B) \cdot \delta_{2,S} + 0 \cdot \delta_{0,S} \qquad (4.7)$$

The dynamics of the system is determined by equation 4.8.

$$\frac{df(E_i)}{dt} = -\frac{(f(E_i) - f_0(E_i))}{\tau} \qquad (4.8)$$

Here $f(E_i)$ and $f_0(E_i)$ are the non-equilibrium and equilibrium occupation probabilities of the levels E_i as indicated in Figure 4.8. The parameters Δ, g, and τ were fit to the experimental data (see Figure 3.27): $\Delta = 40$ meV, $g = 2.7$, $\tau = 1.6$ μ sec. The value of Δ is within the expected range. Evidently the g-factor is increased by ligand interaction. The novel feature is the determination of the dynamical relaxation time τ.

Conclusion

5. Conclusion

The first systematic study of a homologous series of iron(II) spin-crossover coordination polymers containing ClO_4^- and BF_4^- as counteranions are peresented. It is demonstrated that for the $[Fe(ndit z)_3](ClO_4)_2$ and $[Fe(nditz)_3](BF_4)_2$ (n = 4–10 and 12) series both the length and parity of the spacer have a large and systematic influence on the magnetic and photo-magnetic properties. As indicated by XRPD, the phases with $n > 4$ seem to adopt a one dimensional polymer by forming chains, arranged approximately hexagonal close packed. Perchlorate samples in general crystallize more readily than the tetrafluoroborate analogues. The longer the alkylene spacer the higher the disorder of the infinite one dimensional network as indicated by severe reflection broadening The larger iron- iron distances and disorder leads to a decrease in cooperativity between iron centres until the transition can be described by a simple Boltzmann distribution. This behaviour is supported by the magnetic properties presented around the transition points in Figure 5.2. For clarity an insert, which shows the slope of the inflection point of the transition curves, is additionally displayed. Hence it can be seen that the spin transition curves become more gradual with increasing n. Therefore we conclude that the chain arrangement, as it is proposed in the PXRD section, holds for the discussed series.

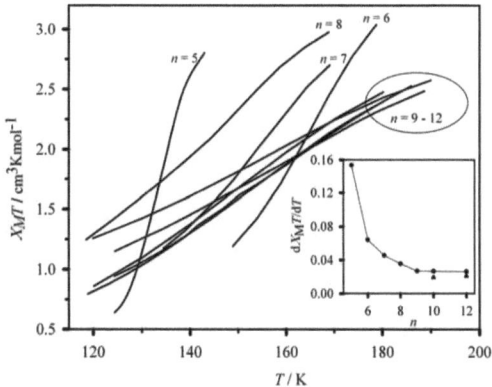

Figure 5.1.: Temperature dependence of $\chi_M T$ for [Fe(nditz)$_3$](BF$_4$)$_2$ with n = 5-10 and 12, [Fe(10ditz)$_3$](ClO$_4$)$_2$ and [Fe(12ditz)$_3$](ClO$_4$)$_2$ in the heating mode close to the transition temperatures. The insert shows the slope of the inflection point of the transition curves. BF$_4^-$ compounds are denoted with (•) [Fe(10ditz)$_3$](ClO$_4$)$_2$ and [Fe(12ditz)$_3$](ClO$_4$)$_2$ are denoted with (▲). The line connecting the points has no physical meaning but is intended as a guide to the eyes.

From the magnetic studies of the [Fe(nditz)$_3$](ClO$_4$)$_2$ series, it is shown that the thermal spin-transition temperature increases with the number of carbon atoms (n) in the spacer and a fascinating effect of the parity is reported (see Figure 4.3). The complexes with an even n display a more abrupt spin transition than odd-numbered ones. The $T_{1/2}$ values have also been found to be higher for the complexes having even-numbered nditz ligands than for the odd-numbered ligands. From the photo-magnetic investigations, it can also be highlighted that the LIESST properties of the [Fe(nditz)$_3$](ClO$_4$)$_2$ family are strongly affected by the parity of the bridging ligands. Based on the SCC model in combination with far-FTIR spectroscopic data, it can be proposed that the energy difference between the LS and the HS states is higher for the even series. Moreover, it is experimentally observed that the level of photo-excitation, as well as the T(LIESST) temperature, perfectly follow the tendency defined by $T_{1/2}$.

By changing the anion to BF_4^-, the structure is slightly changed and thus altered the SC behaviour. In Figure 5.2 $T_{1/2}$ vs. n is shown for the BF_4^- and the ClO_4^- series. The curves show a zigzag behaviour with decreasing amplitude. This suggests that as the alkylene spacer is lengthened $T_{1/2}$ eventually reaches a limiting value of 160 K. It is, however, obvious that the tetrafluoroborates show a limited parity effect from n = 5-7, however if n is increased from 7 to 8 the parity effect is reversed. This suggests that at this length the alkylene chains are differently arranged. Chemical intuition would suggest more or less identical results for such similar non-coordinating anions like perchlorate and tetrafluoroborate. In spite of this the investigations on these two related series show clearly that there is an anion dependence, especially in terms of their magneto-optical behaviour. It is, however impressing that the dependence of the number of carbons n in the spacer can also be found, when analysing the absorption spectra of the complexes of the two series in the UV/VIS-NIR range (see Figure 4.3). Therefore it can be concluded that the distortion of the complexes is linked to n as well as the above mentioned magnetic and magneto-optical properties.

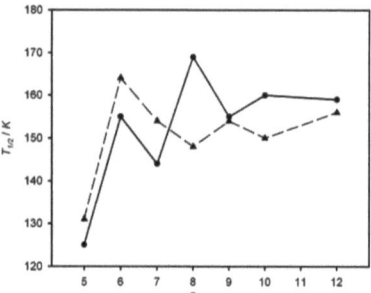

Figure 5.2.: Comparison of $T_{1/2}$ as function of the number of carbons (n) in the spacer of the ditetrazole ligands for the BF_4^- (▲) and ClO_4^- (●) series.

[Fe(4*ditz*)$_3$](BF$_4$)$_2$ and its ClO$_4^-$ analogue are exceptions within the series. [Fe(4*ditz*)$_3$](ClO$_4$)$_2$ shows a two-step transition that is similar to the previously published compound obtained with PF$_6^-$ as anion [52]. A similar feature is obtained in the *T(LIESST)* experiments. The [Fe(4*ditz*)$_3$](BF$_4$)$_2$ shows two interesting effects in its ability (i) to remain even at low temperatures in a mixed HS-LS state and (ii) to remain in a metastable HS state when rapidly cooled. Because of this different properties it can therefore be concluded that the 4*ditz* ligand provides exactly the correct length to allow 3D networks, which is proved through the crystal structure analysis of the [Fe(4*ditz*)$_3$](BF$_4$)$_2$ and [Fe(4*ditz*)$_3$](ClO$_4$)$_2$ complexes, and interesting cooperative effects. Further discussions about the [Fe(4*ditz*)$_3$](X)$_2$ complexes can be found in [62].

Investigations of the magnetic field induced HS/LS transition of the [Fe(4*ditz*)$_3$](PF$_6$)$_2$ complex in transient megagauss fields using optical reflection measurements at λ=0.632 nm leads to induced small reflection changes with strong hysteresis. Evidently the strong (160 T) but short (5 μsec) magnetic field pulse cannot drive the transition into saturation. Using magnetic field pulses with different time constants the dynamic process could be studied.

To simulate the experimental results the model of a split S=2/S=0 level system was applied. The parameters Δ, g, and τ were fit to the experimental data: Δ = 40 meV, g = 2.7, τ = 1.6 μsec. The value of Δ is within the expected range. Evidently the g factor is increased by ligand interaction. The determination of the dynamical relaxation time τ is presented for the first time. The development of the device, provides the opportunity to induce the SC at fields higher than 100 T. Therefore it is now possible to measure SC systems, which do not show any reaction in fields smaller than this.

Appendix

A. Reflectivity

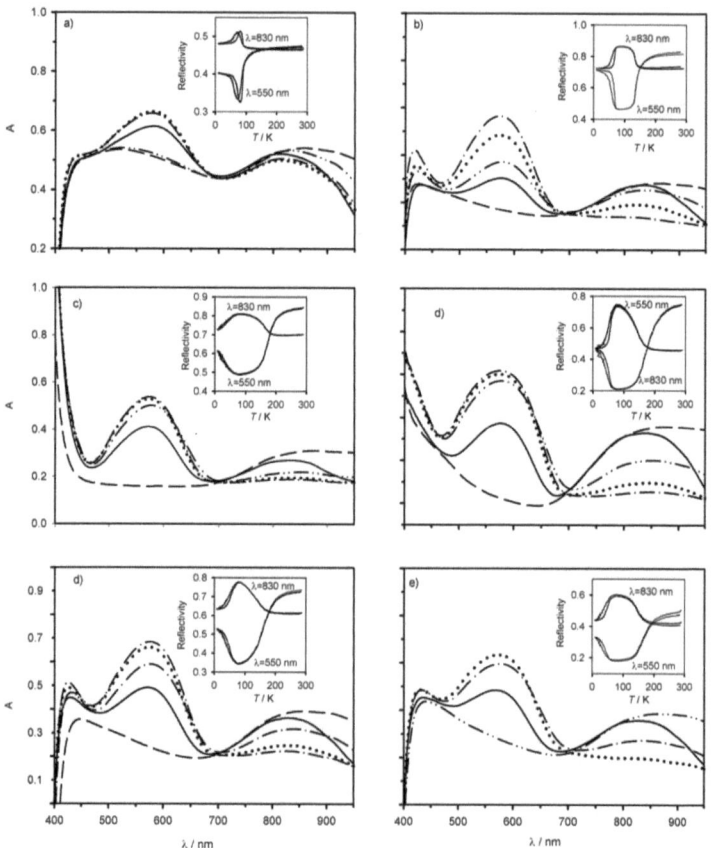

Figure A.1.: Reflectivity measurements of the [Fe(n*ditz*)$_3$](BF$_4$)$_2$ complexes with n = 4-9.

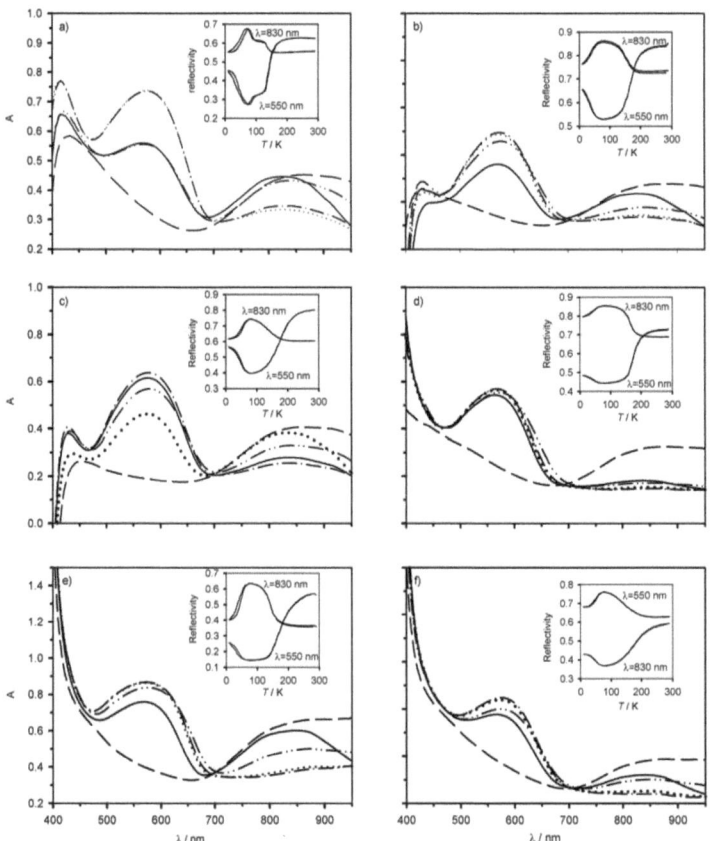

Figure A.2.: Reflcetivity measurements of the [Fe(n*ditz*)₃](ClO₄)₂ complexes with n = 4,6-9 and 12.

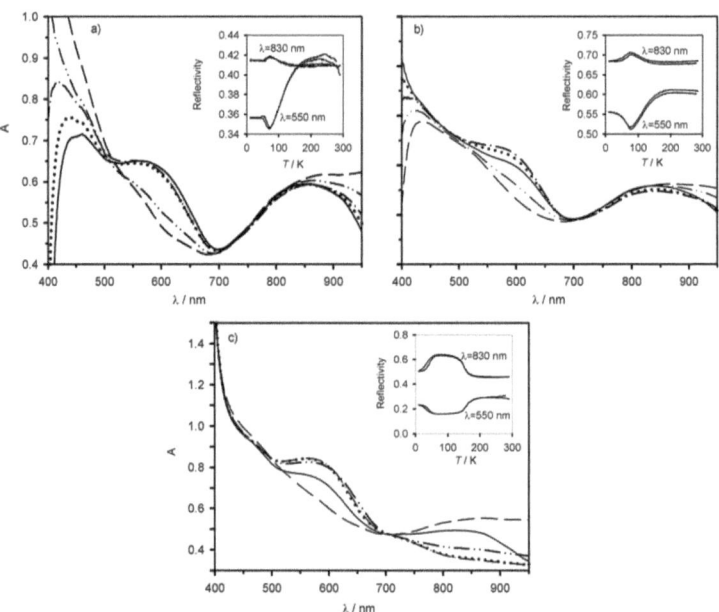

Figure A.3.: Reflectivity measurements of the [Fe(n*ditz*)$_3$](PF$_6$)$_2$ complexes with n = 7-9.

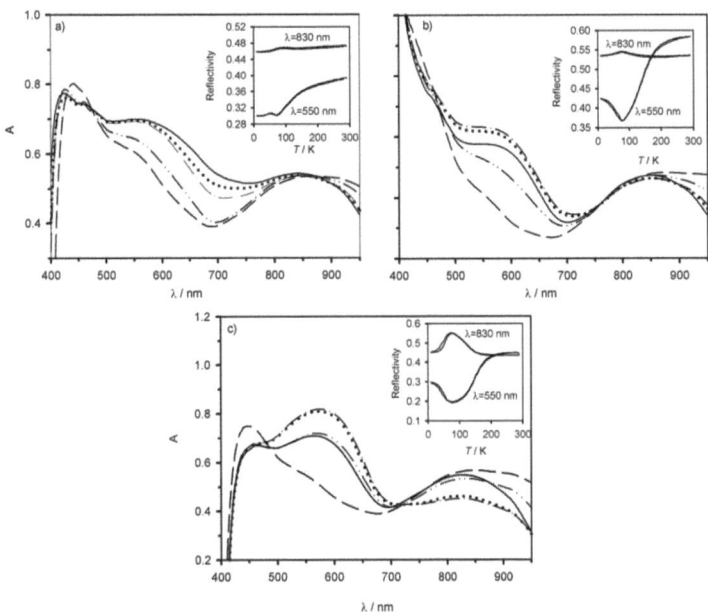

Figure A.4.: Reflectivity measurements of the [Fe(n*ditz*)₃](SbF₆)₂ complexes with $n = 7\text{-}9$.

B. Publications

- **Both Spacer Length and Parity Influence the Thermal and Light-Induced Properties of Iron(ii) α,ω-Bis(tetrazole-1-yl)alkane Coordination Polymers**
  Alina Absmeier, Matthias Bartel, Chiara Carbonera, Guy N. L. Jameson, Peter Weinberger, Andrea Caneschi, Kurt Mereiter, Jean-Francois Létard, and Wolfgang Linert

- **Field induced low-spin high-spin transition**
  B. Richtera, A. Kirstea, S. Hansela, M. von Ortenberga, A. Absmeier, W. Linert, R. Groessinger

FULL PAPER

DOI: 10.1002/chem.200500941

Both Spacer Length and Parity Influence the Thermal and Light-Induced Properties of Iron(II) α,ω-Bis(tetrazole-1-yl)alkane Coordination Polymers

Alina Absmeier,[a, c] Matthias Bartel,[a, b] Chiara Carbonera,[c] Guy N. L. Jameson,[a] Peter Weinberger,[a] Andrea Caneschi,[b] Kurt Mereiter,[d] Jean-François Létard,[c] and Wolfgang Linert*[a]

Abstract: A new series of [μ-tris-{1,n-bis(tetrazol-1-yl)alkane-N4,N4′}iron(II)] bis(perchlorate) spin-crossover coordination polymers ([Fe(nditz)$_3$](ClO$_4$)$_2$]; n=4–9) has been synthesised and characterised. The ditetrazole bridging ligands provide octahedral symmetry at the iron(II) centres while allowing the distance between iron(II) centres to be varied. These polymers have therefore been investigated to determine the effects of spacer length on their thermal and light-induced spin-transition behaviour. An increase in the number of carbon atoms in the spacer (n) raises the thermal spin-crossover temperature, while decreasing the stability of the light-induced metastable state generated through the light-induced excited spin state trapping (LIESST) effect by irradiating the sample at 530 nm. Remarkably, however, the parity of the spacer also has an effect, enabling the series of complexes to be divided into two sub-series depending on whether the bridging ligand possesses an even or an odd number of carbon atoms. An explanation at the molecular level using the single configurational coordinate (SCC) model is presented.

Keywords: iron · magnetic properties · nitrogen heterocycles · organic–inorganic hybrid composites · photo-magnetism · spin crossover

Introduction

Compounds that are able to switch their magnetic properties through changes in external conditions (e.g. temperature, pressure, light, etc.) have been studied for many years[1,2] because of their potential application in data-storage systems. In such systems, optical switching is particularly promising

[a] A. Absmeier, M. Bartel, Dr. G. N. L. Jameson, Dr. P. Weinberger, Prof. Dr. W. Linert
Institute of Applied Synthetic Chemistry
Vienna University of Technology
Getreidemarkt 9/163-AC, 1060 Vienna (Austria)
Fax: (+43) 1-58801-16299
E-mail: wlinert@mail.zserv.tuwien.ac.at

[b] M. Bartel, Prof. Dr. A. Caneschi
LAMM, Dipartimento di Chimica & UdR INSTM
Università di Firenze
Via della Lastruccia 3, 50019 Sesto F.no (Italy)

[c] A. Absmeier, Dr. C. Carbonera, Prof. Dr. J.-F. Létard
Institut de Chimie de la Matière Condensée de Bordeaux
UPR CNRS No 9048, Université Bordeaux 1
Groupe des Sciences Moléculaires
87 Av. du Doc. A. Schweitzer, 33608 Pessac (France)

[d] Prof. Dr. K. Mereiter
Institute for Chemical Technologies and Analytics
Vienna University of Technology
Getreidemarkt 9/164-SC, 1060 Vienna (Austria)

because of the high speed and ease of writing and reading information. This mode of switching thermally induced spin-crossover (SC) compounds was made possible following the discovery of light-induced excited spin state trapping (LIESST), which allows the spin state of an iron(II) centre to be changed from low spin (LS) to high spin (HS) and then restored through the reverse LIESST phenomenon.[3,4] The major drawback of this process is that although the lifetime of the photo-induced HS state is almost infinite below approximately 50 K, at higher temperatures the decay increases rapidly, thus making data storage impossible. It is therefore important to identify and understand the nature of the parameters affecting the stability of the photo-induced state,[5,6] and indeed many different complexes have been studied for the LIESST effect. Of those containing iron(II), however, the majority use terminal ligands[3,7–9] and are thus only able to form mononuclear complexes. Notable exceptions involve dimers[10,11] and a mixed-metal bridging system with a 2D layered structure.[12] Furthermore, a few systems have been investigated with a homologous series of ligands, which would allow a more systematic approach to the study of the various influences of the ligand.

Iron(II) spin-crossover compounds have been produced with many different N-coordinating heterocyclic ligands, such as triazoles,[13] tetrazoles[3,7–9,14] and imidazoles.[15] In particular, tetrazole-containing ligands have been successfully employed as part of a family[16] of terminal and bridging ligands to give mononuclear complexes[3,7–9] and within bridging ligands to form 1D chain structures[17,18] and 3D networks.[14] The latter, [μ-tris-{1,4-bis(tetrazol-1-yl)butane-N4,N4'}iron(II)] bis(hexafluorophosphate) ([Fe(4ditz)₃]-(PF₆)₂), proved to be particularly interesting. Each iron(II) centre is octahedrally coordinated by symmetry-equivalent tetrazole rings with the butylene spacer outstretched, thereby assuming a zigzag *trans* configuration spanning iron(II) centres (see Figure 1). The ditetrazole ligands link the iron(II) atoms into a 3D network, and three such 3D net-

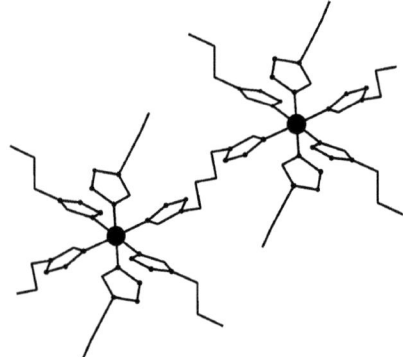

Figure 1. The basic structural units of [Fe(4ditz)₃](PF₆)₂ showing how the ligands span the iron(II) centres to produce a three-dimensional polymer.

works interpenetrate each other. This structure results in cooperativity between the iron(II) centres, as borne out by SQUID measurements. The compound has a two-step spin transition at 168 K (with a hysteresis of 5 K) and 173 K.[14] While the shorter ethylene-bridged ligand 1,2-bis(tetrazol-1-yl)ethane produces a 1D polymeric chain structure,[18] it was thought possible that longer alkyl spacers (n > 4, where n is the number of carbon atoms in the spacer) might allow more interlocking networks and therefore exhibit interesting new effects.

We present here a detailed study of a series of α,ω-bis(tetrazol-1-yl)alkane iron(II) bis(perchlorate) coordination polymers in which the alkane is varied in the series from butane to nonane (n = 4–9). All compounds have been characterised by UV/Vis-NIR and FTIR spectroscopy, reflectivity experiments, SQUID and LIESST. The effect of the spacer length and also the parity of the spacer on the thermal and light-induced properties in particular are discussed.

Experimental Section

Chemicals and standard physical characterisation: L-Ascorbic acid, 1,5-diaminopentane (>98%), 1,7-diaminoheptane (99%), 1,9-diaminononane (99%), glacial acetic acid (99%), iron(II) perchlorate hexahydrate, sodium azide, sodium hydroxide (97%) and triethyl orthoformate were obtained from Aldrich. All other chemicals were standard reagent grade and used as supplied.
Elemental analyses (C, H and N) were performed by the Mikroanalytisches Laboratorium, Faculty of Chemistry, Vienna University, Währingerstrasse 42, 1090 Vienna, Austria. ¹H and ¹³C NMR spectra in [D₆]DMSO were measured with Bruker DPX 200 MHz or Bruker 250 FS FT-NMR spectrometers. ¹H NMR chemical shifts are reported in ppm calibrated to the respective solvent. Mid-range FTIR spectra of the compounds were recorded as KBr pellets in the range 4400–450 cm⁻¹ with a Perkin-Elmer 16PC FTIR spectrometer. Pellets were obtained by pressing a powdered mixture of the samples in KBr in vacuo in a hydraulic press at a pressure of 10000 kgcm⁻² for 5 min. Far-range FTIR spectra were recorded in the range 600–250 cm⁻¹ on a Perkin–Elmer System 2000 Far-FTIR spectrometer. The complexes were diluted with polyethylene and pressed at a pressure of 10000 kgcm⁻² transiently. Variable-temperature far-IR spectra in the temperature range 100–298 K were recorded with a Graseby-Specac thermostattable sample holder with polyethylene windows, attached to a Graseby-Specac automatic temperature controller.

Synthesis and crystallographic studies of the ligands: The general synthetic pathway and the synthesis of the even-numbered ligands 4ditz, 6ditz and 8ditz have been reported in the literature.[19,20] Modified procedures were used to produce three odd-numbered ligands 1,5-bis(tetrazol-1-yl)pentane (5ditz), 1,7-bis(tetrazol-1-yl)heptane (7ditz) and 1,9-bis(tetrazol-1-yl)nonane (9ditz), as described below.

The respective diamine (80 mmol), sodium azide (160 mmol) and triethyl orthoformate (160 mmol) were stirred in a 500-mL, three-necked, round-bottomed flask. Acetic acid (250 mL, 99.5%) was then added and the mixture heated to 90–95 °C for four hours. After 4 and 16 h reaction time another aliquot of triethyl orthoformate (160 mmol) and sodium azide (160 mmol) was added and the mixture stirred for an additional 24 h at 95 °C. After cooling, the reaction mixture was poured into a beaker and a saturated sodium hydrogencarbonate solution was added with vigorous stirring to neutralise the acetic acid, followed by solid sodium hydrogencarbonate to precipitate the product. The suspension was cooled to 4 °C for 3 h and the precipitate was filtered off and recrystallised from ethanol. The colourless needle-shaped crystals were dried over P₂O₅. For all three compounds the mid-FTIR spectra show prominent absorptions at 3115 cm⁻¹ (ν_{C-H} of the aromatic tetrazole ring); 2950/2940/2939 and 2870/2865/2849 cm⁻¹ (ν_{C-H} of the aliphatic C–H in the pentylene/heptylene/nonylene spacer); and 1790/1792/1793, 1490/1491/1492, 1460/1464/1463 and 1175/1174 cm⁻¹ (typical ν_{C-C} and ν_{C-N} of the tetrazole rings in 5ditz/7ditz/9ditz). The single crystals used for X-ray diffraction were obtained by solvent evaporation from an ethanol solution (5ditz), a pyridine solution (7ditz) or a DMF solution (9ditz).

5ditz: Yield: 10%; m.p. 125–127 °C; ¹H NMR (250 MHz, [D₆]DMSO): δ = 9.39 (s, 2H), 4.44 (t, J = 7.03 Hz, 4H), 1.86 (quin, J = 7.19 Hz, 4H), 1.18 ppm (quin, J = 7.62 Hz, 2H); ¹³C NMR (50 MHz, [D₆]DMSO): δ = 143.8 (d, 2C), 47.1 (t, 2C), 28.3 (t, 2C), 22.4 ppm (t, 2C); elemental analysis calcd (%) for C₇H₁₂N₈: C 40.38, H 5.81, N 53.81; found: C 40.65, H 5.72, N 53.52.

7ditz: Yield: 6.4%; m.p. 85–86 °C; ¹H NMR (200 MHz, [D₆]DMSO): δ = 9.38 (s, 2H), 4.43 (t, J = 7.05 Hz, 4H), 1.81 (quin, J = 7.09 Hz, 4H), 1.22 ppm (m, 6H); ¹³C NMR (50 MHz, [D₆]DMSO): δ = 143.7 (d, 2C), 47.3 (t, 2C), 28.9 (t, 2C), 27.4 (t, 2C), 25.4 ppm (t, 2C); elemental analysis calcd (%) for C₉H₁₆N₈: C 45.75, H 6.83, N 47.42; found: C 45.97, H 6.90, N 47.32.

9ditz: Yield: 22.3%; m.p. 92–93 °C; ¹H NMR (200 MHz, [D₆]DMSO): δ = 9.39 (s, 2H), 4.43 (t, J = 7.14 Hz, 4H), 1.81 (quin, J = 7.17 Hz, 4H), 1.21 ppm (m,10H); ¹³C NMR (50 MHz, [D₆]DMSO): δ = 143.7 (d, 2C), 47.4 (t, 2C), 29.0 (t, 2C), 28.4 (t), 28.0 (t, 2C), 25.5 ppm (t, 2C); elemental analysis calcd (%) for C₁₁H₂₀N₈: C 49.98, H 7.63, N 42.39; found: C 50.06, H 7.73, N 42.32.

Crystals of 5ditz, 7ditz and 9ditz were all elongated, lath-like and soft. Selected crystals were mounted on a Bruker SMART diffractometer (graphite-monochromated Mo$_{Kα}$ radiation from a sealed X-ray tube; λ =

0.71073 Å, platform three-circle goniometer, CCD area detector) and intensity data were collected at room temperature. After raw data extraction with the program SAINT, absorption and related effects were corrected with the program SADABS (multi-scan method) and data were processed with XPREP.[21] The structures were then solved by direct methods using SHELXS-97 followed by structure refinements on F^2 with SHELXL-97.[22] Non-hydrogen atoms were refined anisotropically. Hydrogen atoms were inserted in calculated positions and refined with the riding model. Crystallographic data are given in Table 1 and ORTEP plots are given in Figure 2 with the corresponding even ligands. CCDC-268788 (5ditz), CCDC-268789 (7ditz), and CCDC-268790 (9ditz) contain the supplementary crystallographic data for this paper. These data can be obtained free of charge from the Cambridge Crystallographic Data Centre via www.ccdc.cam.ac.uk/data_request/cif.

Synthesis of the complexes: A similar procedure was used for the synthesis of all members of the series [Fe(nditz)$_3$](ClO$_4$)$_2$ (n = 4–9). The respective ligand (1 mmol) was dissolved in hot ethanol. While the solution cooled down to 40 °C, iron(II) perchlorate hexahydrate (0.33 mmol) and a small amount of ascorbic acid (to keep the iron as iron(II)) were diluted in ethanol (5 mL). This solution was slowly added to the dissolved ligand and the resulting mixture stirred for four hours. The precipitate was filtered off and the obtained powder dried over P$_2$O$_5$. Unfortunately, it has so far proved impossible to grow single crystals by any of the standard methods, including H-tube slow diffusion or slow cooling.

Elemental analyses and mid-FTIR data

[Fe(4ditz)$_3$](ClO$_4$)$_2$: Yield: 80%. Elemental analysis calcd (%) for C$_{18}$H$_{30}$Cl$_2$FeN$_{24}$O$_8$: C 25.82, H 3.61, N 40.15; found: C 26.59, H 3.58, N 39.36. Mid-FTIR: $\tilde{\nu}$ = 3137 ($\nu_{\text{C1-H1}}$ of the aromatic tetrazole ring); 2956 and 2875 ($\nu_{\text{C-H}}$ of the aliphatic C–H in the butylene spacer); 1784, 1651, 1470, 1445 and 1183 cm^{-1} (typical $\nu_{\text{C-C}}$ and $\nu_{\text{C-N}}$ of the tetrazole rings).

[Fe(5ditz)$_3$](ClO$_4$)$_2$: Yield: 82%. Elemental analysis calcd (%) for C$_{21}$H$_{36}$Cl$_2$FeN$_{24}$O$_8$: C 28.68, H 4.13, N 38.23; found: C 29.65, H 4.03, N 37.53. Mid-FTIR: $\tilde{\nu}$ = 3135 ($\nu_{\text{C1-H1}}$ of the aromatic tetrazole ring); 2947 and 2866 ($\nu_{\text{C-H}}$ of the aliphatic C–H in the pentylene spacer); 1787, 1649, 1459, 1445 and 1181 cm^{-1} (typical $\nu_{\text{C-C}}$ and $\nu_{\text{C-N}}$ of the tetrazole rings).

[Fe(6ditz)$_3$](ClO$_4$)$_2$: Yield: 76%. Elemental analysis calcd (%) for C$_{24}$H$_{42}$Cl$_2$FeN$_{24}$O$_8$: C 31.28, H 4.59, N 36.48; found: C 31.24, H 4.45, N 35.68. Mid-FTIR: $\tilde{\nu}$ = 3137 ($\nu_{\text{C1-H1}}$ of the aromatic tetrazole ring); 2935 and 2860 ($\nu_{\text{C-H}}$ of the aliphatic C–H in the hexylene spacer); 1781, 1636, 1452, 1438 and 1178 cm^{-1} (typical $\nu_{\text{C-C}}$ and $\nu_{\text{C-N}}$ of the tetrazole rings).

[Fe(7ditz)$_3$](ClO$_4$)$_2$: Yield: 43%. Elemental analysis calcd (%) for C$_{27}$H$_{48}$Cl$_2$FeN$_{24}$O$_8$: C 33.66, H 5.02, N 34.69; found: C 33.78, H 4.84, N 34.53. Mid-FTIR: $\tilde{\nu}$ = 3137 ($\nu_{\text{C1-H1}}$ of

Table 1. Crystallographic data of 5ditz, 7ditz and 9ditz.

	5ditz	7ditz	9ditz
formula	C$_7$H$_{12}$N$_8$	C$_9$H$_{16}$N$_8$	C$_{11}$H$_{20}$N$_8$
molecular weight	208.25	236.30	264.35
crystal size [mm]	0.60 × 0.22 × 0.20	1.00 × 0.25 × 0.04	1.00 × 0.25 × 0.04
space group	Fdd2 (no. 43)	Fdd2 (no. 43)	Fdd2 (no. 43)
a [Å]	14.111(2)	13.738(2)	13.393(4)
b [Å]	32.041(4)	38.770(7)	45.260(13)
c [Å]	4.5613(6)	4.6300(8)	4.6724(14)
V [Å^3]	2062.3(5)	2466.0(7)	2832.4(15)
Z	8	8	8
ρ_{calcd} [g cm^{-3}]	1.341	1.273	1.240
T [K]	297(2)	297(2)	297(2)
μ [mm^{-1}] (MoK_α)	0.095	0.088	0.084
F(000)	880	1008	1136
θ_{max} [°]	25	30	25
no. of reflections measured	4273	8879	6658
no. of unique reflections	912	1788	1240
no. of reflections $I > 2\sigma(I)$	885	1365	984
no. of parameters	69	78	87
R_1 ($I > 2\sigma(I)$)[a]	0.0444	0.0391	0.0547
R_1 (all data)	0.0462	0.0558	0.0726
wR_2 (all data)	0.1083	0.1069	0.1549
difference Fourier peaks min./max. [e Å^{-3}]	−0.11/0.12	−0.10/0.12	−0.13/0.22

[a] $R_1 = \Sigma ||F_o| - |F_c|| / \Sigma |F_o|$, $wR_2 = [\Sigma(w(F_o^2 - F_c^2)^2) / \Sigma(w(F_o^2)^2)]^{1/2}$.

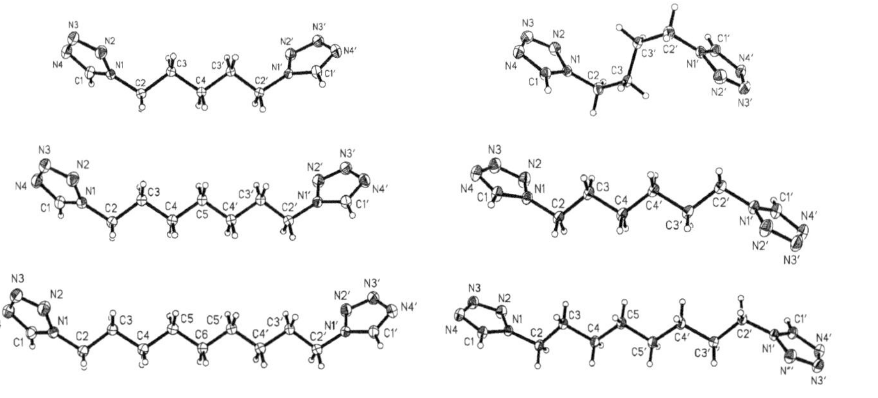

Figure 2. Structural views of 5ditz, 7ditz and 9ditz (20% probability ellipsoids). All molecules have C_2 symmetry with the twofold axis passing through C4, C5 and C6, respectively. 4ditz, 6ditz and 8ditz are shown to the right for comparison.[16]

the aromatic tetrazole ring); 2940 and 2864 (v_{C-H} of the aliphatic C–H in the heptylene spacer); 1790, 1635, 1463, 1443 and 1180 cm^{-1} (typical v_{C-C} and v_{C-N} of the tetrazole rings).

[Fe(8ditz)$_3$](ClO$_4$)$_2$]: Yield: 96%. Elemental analysis calcd (%) for C$_{30}$H$_{54}$Cl$_2$FeN$_{24}$O$_8$: C 35.83, H 5.41, N 33.43; found: C 36.64, H 5.16, N 32.88. Mid-FTIR: $\tilde{v} = 3138$ (v_{Cl-H} of the aromatic tetrazole ring); 2934 and 2859 (v_{C-H} of the aliphatic C–H in the octylene spacer); 1792, 1645, 1457, 1440 and 1181 cm^{-1} (typical v_{C-C} and v_{C-N} of the tetrazole rings)

[Fe(9ditz)$_3$](ClO$_4$)$_2$]: Yield: 86%. Elemental analysis calcd (%) for C$_{33}$H$_{60}$Cl$_2$FeN$_{24}$O$_8$: C 37.83, H 5.77, N 32.08; found: C 38.30, H 5.54, N 30.84. Mid-FTIR: $\tilde{v} = 3138$ (v_{Cl-H} of the aromatic tetrazole ring); 2929 and 2856 (v_{C-H} of the aliphatic C–H in the nonylene spacer); 1798, 1650, 1462, 1444 and 1180 cm^{-1} (typical v_{C-C} and v_{C-N} of the tetrazole rings).

UV/Vis–NIR reflectivity: UV/Vis–NIR spectra were recorded with a Perkin–Elmer Lambda 900 UV/Vis–NIR spectrometer between 1500 and 300 nm using the method of diffuse reflection. A spectrum of BaSO$_4$ was subtracted as background. Variable-temperature measurements were made using a custom-made thermostattable sample holder with quartz glass windows within a spectralon integration sphere. The temperature was controlled with a Harrick controller. Aluminium foil was used to improve the thermal contact between the sample holder and the sample. The spectra were measured between 105 and 260 K in intervals of 5 to 10 K.

The reflectivity of the samples was further investigated with a custom-built reflectivity set-up equipped with a CVI spectrometer, which allows the collection of both the reflectivity spectra within the range of 450–950 nm at a given temperature and to follow the temperature dependence of the signal at a selected wavelength (±2.5 nm) at 5–290 K. The analysis was performed on a thin layer of the powdered sample without any dispersion in a matrix.[23]

Magnetic susceptibility and magneto-optical measurements: Magnetic measurements were recorded on two SQUID (Superconducting Quantum Interference Device) magnetometers as described hereafter: 1) SQUID Cryogenix S600 magnetometer with an applied field of 1 T; 2) MPMS-55 Quantum Design SQUID magnetometer with an operating field of 2 T within the temperature range of 2–300 K and with a speed of 10 K min^{-1} in the settle mode at atmospheric pressure. All measurements were performed on polycrystalline powder samples weighing about 12 mg. The data were corrected for the magnetisation of the sample holder and for diamagnetic contributions, estimated from Pascal's constants.

The photo-magnetic measurements were performed with a Spectra Physics Series 2025 Kr$^+$ laser ($\lambda = 532$ nm) coupled by an optical fibre to the cavity of the SQUID magnetometer (MPMS-55 Quantum Design SQUID) operating with an external magnetic field of 2 T within the 2–300 K temperature range and a speed of 10 K min^{-1} in the settle mode at atmospheric pressure. The power at the sample was adjusted to 5 mW cm^{-2}. Bulk attenuation of light intensity was limited as much as possible by the preparation of a thin layer of compound. It is noteworthy that there was no change in the data due to sample heating upon laser irradiation. The weight of these thin layer samples (approximately 0.2 mg) was obtained by comparison of the measured thermal spin-crossover curve with another curve of a more accurately weighed sample of the same compound.

Results and Discussion

The coordination polymers described herein form a series that differ only in the length of the alkane spacer connecting the tetrazole moieties. Unfortunately, it was not possible, despite considerable effort, to produce single crystals suitable for X-ray diffraction studies. However, some general comments about the structure of these polymers can be made. Firstly, the structures can be compared with that of [Fe(4ditz)$_3$](PF$_6$)$_2$, which was published previously[19] and is described above (see also Figure 1). Preliminary powder X-ray diffraction investigations of the ClO$_4^-$ and PF$_6^-$ salts of [Fe(4ditz)$_3$] show that they have a similar structure, with ClO$_4^-$ having reduced symmetry due to compression along one or more axes. As the group is traversed, the anion and the method of production are kept constant, with only the alkane spacer changing. The crystal packing of all the free ligands is very similar, thereby suggesting a similar behaviour. Furthermore, the UV/Vis-IR data presented below show that the iron is octahedrally coordinated. Thus, although the presence or number of interpenetrating networks seen in [Fe(4ditz)$_3$](PF$_6$)$_2$ is unknown, it can be assumed that a 3D network of octahedrally coordinated iron centres, as shown in Figure 1, is present in all samples. Naturally, as the alkane spacer increases in length, the increase in degrees of freedom reduces the crystallinity of the samples (i.e. making them more amorphous). However, because the iron centres themselves are well defined, a non-crystalline structure in no way precludes important information being gained from measurement of the magnetic properties.

Magnetic properties: To study how the number of carbon atoms in the spacer of the ligand influences the spin transition, susceptibility curves were recorded between 10 and 300 K for all the complexes of the series. Figure 3 shows the obtained $\chi_M T$ versus T curves, where χ_M is the molar magnetic susceptibility and T is the temperature.

All the investigated products undergo a thermal spin-transition at around 150 K, from a $\chi_M T$ product close to 3.0 cm^{-3} K mol^{-1} at room temperature, in agreement with the expected HS state, and a diamagnetic value at low temperature reflecting the LS state. In some of the curves a residual magnetisation is observed at low temperatures, yielding a $\chi_M T$ value of up to 1 cm^{-3} K mol^{-1}, which might be caused by traces of iron(III) or thermally SC-inactive HS iron(II).

The temperature of the thermal spin-state transition, $T_{1/2}$, estimated from the maximum in the derivative of the $\chi_M T$ versus T plot is given in Table 2, and Figure 4a shows the change of $T_{1/2}$ with the length of the alkane spacer (n). Interestingly, an increase in the length of the spacer raises the $T_{1/2}$ value, but it can also be seen that the influence of the parity of the spacer is not negligible. Indeed, it seems that the complexes can be divided into two series depending on the parity of the spacer: the first, which contains bridging ligand with odd-numbered carbon atoms in the spacer, shows a $\chi_M T$ product at room temperature that is always equal to 3.0 cm^{-3} K mol^{-1}; the second series, with even-numbered spacers, exhibits a $\chi_M T$ at room temperature of approximately 3.75 cm^{-3} K mol^{-1}. A similar classification can also be made if we compare the shape of the thermal spin-transitions in Figure 3. Complexes with an odd-numbered spacer ([Fe(5ditz)$_3$](ClO$_4$)$_2$, [Fe(7ditz)$_3$](ClO$_4$)$_2$ and [Fe(9ditz)$_3$](ClO$_4$)$_2$]) display a more gradual transition than complexes with an even-numbered ligand ([Fe(4ditz)$_3$](ClO$_4$)$_2$, [Fe(6ditz)$_3$](ClO$_4$)$_2$ and [Fe(8ditz)$_3$](ClO$_4$)$_2$).

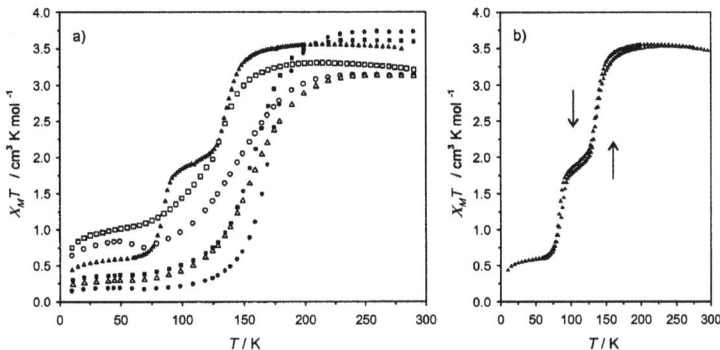

Figure 3. a) Temperature dependency of the $\chi_M T$ product for the investigated complexes [Fe(4ditz)$_3$](ClO$_4$)$_2$ (▲), [Fe(5ditz)$_3$](ClO$_4$)$_2$ (□), [Fe(6ditz)$_3$]-(ClO$_4$)$_2$ (■), [Fe(7ditz)$_3$](ClO$_4$)$_2$ (○), [Fe(8ditz)$_3$](ClO$_4$)$_2$ (●) and [Fe(9ditz)$_3$](ClO$_4$)$_2$ (△); b) cooling and heating modes for [Fe(4ditz)$_3$](ClO$_4$)$_2$.

Table 2. Spin-transition, reflectivity and LIESST parameters of the discussed complexes.

[Fe(nditz)$_3$](ClO$_4$)$_2$	$T_{1/2}$ [K]	T(LIESST) [K]	% Irr-surface[b]	% Irr-bulk[c]
[Fe(4ditz)$_3$](ClO$_4$)$_2$	84/134[a]	58/39	52	60
[Fe(5ditz)$_3$](ClO$_4$)$_2$	125	52	64	72
[Fe(6ditz)$_3$](ClO$_4$)$_2$	155	–	42	43
[Fe(7ditz)$_3$](ClO$_4$)$_2$	144	51	42	57
[Fe(8ditz)$_3$](ClO$_4$)$_2$	169	–	14	13
[Fe(9ditz)$_3$](ClO$_4$)$_2$	155	37	26	47

[a] Estimated at the maximum of the derivative of each of the two thermal spin-transitions. [b] Percentage of sample photo-bleached. [c] Percentage conversion to the metastable high-spin state by irradiation at 10 K.

However, within this classification the case of [Fe(4ditz)$_3$](ClO$_4$)$_2$ is somewhat exceptional in that it is the only one that displays a two-step thermal spin-transition with a plateau between 93 and 124 K (Figure 3a). This behaviour can be compared with the same complex synthesised in methanol solution and published previously by van Koningsbruggen et al.[19] The magnetic susceptibility curves of the two [Fe(4ditz)$_3$](ClO$_4$)$_2$ compounds (one synthesised in methanol and the other synthesised in ethanol) show clear differences in both the shape of the spin-transition curves and in their spin-transition behaviour. Indeed, the shape of the magnetic susceptibility curve of the compound made previously in methanol shows an incomplete one-step spin transition. This suggests a significant influence of the solvent and could be due to either inclusion of solvent, as observed previously in the [Fe(2-pic)$_3$]Cl$_2$·solvent family of complexes,[24] or the production of different polymorphs according to the solvent. The effect of solvent on the present coordination polymers is currently being investigated further.

Optical and far-FTIR properties: As expected, all members of the series show a thermochromic effect associated with

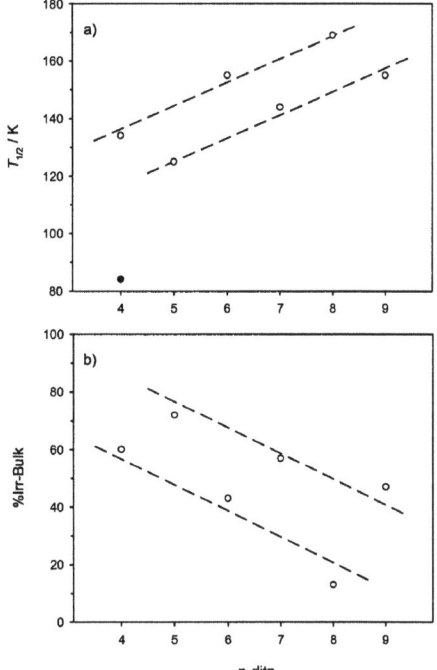

Figure 4. $T_{1/2}$ (a) and %Irr-Bulk (b) as a function of the number of carbons (n) in the spacer of the ditetrazole ligands. (●) indicates $T_{1/2}$ of the lower temperature spin-transition of [Fe(4ditz)$_3$](ClO$_4$)$_2$. The parallel dashed lines are drawn to clearly show the separation of the data points into two sub-series (even and odd) but have no further significance.

the spin transition, from white in the HS state to violet in the LS state. A typical temperature-dependent UV/VIS–NIR spectrum of these complexes is given in Figure 5. The spectrum shown is that of [Fe(9ditz)$_3$](ClO$_4$)$_2$ within the range 300–1500 nm and at temperatures between 105 and 256 K. The absorption spectrum is composed of one charge-transfer (CT) band at low wavelength and three recognisable peaks, two of which decrease and one that increases with increasing temperature. The peaks at 550 and 370 nm are d–d transitions of the LS state of the complex, the peak at 300 nm is a charge-transfer band of the ligand and the band at 1000 nm is the d–d transition of the HS state. According to the Tanabe–Sugano diagram for d^6 systems, the two LS transitions can be assigned to the spin-allowed transitions $^1A_1 \rightarrow {}^1T_2$ and $^1A_1 \rightarrow {}^1T_1$ and the HS band corresponds to the $^5T_2 \rightarrow {}^5E$ transition.

In parallel to this we used a custom-built reflectivity instrument to record the change occurring at 830±2.5 nm and at 550±2.5 nm, which are the wavelengths of the $^5T_2 \rightarrow {}^5E$ and the $^1A_1 \rightarrow {}^1T_1$ transitions, respectively, as a function of temperature (290–10 K) and also light irradiation. The results of a typical experiment are shown in Figure 6 for [Fe(5ditz)$_3$](ClO$_4$)$_2$.

In agreement with the previous UV-VIS–NIR study, within the range of 450–950 nm only the HS band is visible at room temperature for all compounds, whilst at 80 K only the LS transition is observable. Below 80 K, however, the reflectivity experiment demonstrates the existence of a photo-induced phenomenon at the surface. In fact, at low temperatures the sample was seen to bleach (see, for example, Figure 6 and insert), which indicates the occurrence of a LS/HS photo-conversion through the LIESST effect. We have calculated the amount of photo-bleached fraction (%Irr-Surface; reported in Table 2) for each compound rela-

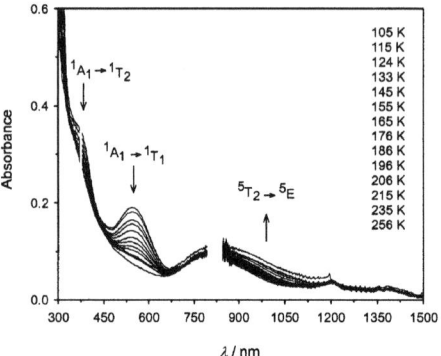

Figure 5. Temperature-dependent UV/VIS–NIR spectra of [Fe(9ditz)$_3$](ClO$_4$)$_2$ between 1500 and 300 nm. The gaps in the data occur because we have removed the large noise caused by changing grating and detector.

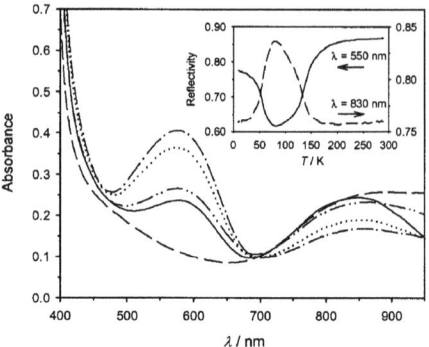

Figure 6. Reflectivity measurements of [Fe(5ditz)$_3$](ClO$_4$)$_2$ as a function of temperature. The absorbance spectra at 280 (---), 140 (-----), 80 (-·-·-), 60 (·····) and 10 K (——) are shown in the main figure. The insert reports the reflectivity followed at 550±2.5 nm and 830±2.5 nm.

tive to the value found at room temperature. Interestingly, the level of photo-excitation roughly decreases with the length of the spacer.

It cannot be excluded that other parameters, such as the morphology of the powder particles, should be taken into account in order to discuss the problem of the light penetration, but it is clear that the level of photo-excitation follows the increase of the $T_{1/2}$ value: the higher the $T_{1/2}$, the lower is the level of the light-induced LS/HS conversion. This result can, in fact, be discussed in terms of the inverse energy-gap law introduced by Hauser.[25] In the so-called single configurational coordinate (SCC) model, the potential wells of the LS and HS states are plotted along a single reaction coordinate Q, which describes the totally symmetric breathing mode and is related to the metal–ligand bond difference by $\Delta Q = \sqrt{6}\Delta r$ (see Figure 7). The potential wells and their shape can, therefore, be moved horizontally relative to each other by changing the relative bond strengths of HS and LS, or the wells can be moved vertically relative to each other.

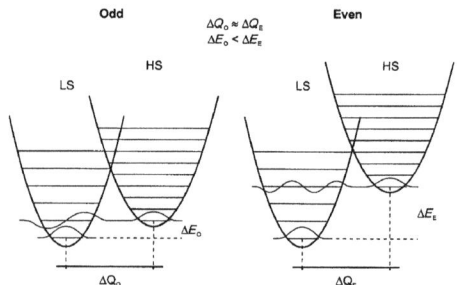

Figure 7. Potential wells for an odd and even ditetrazole coordination polymer demonstrating the effect of the parity of the bridging ligand. Only ΔE appears to be affected and not ΔQ.

The difference in the zero-point energies of LS and HS, ΔE, is therefore related to $T_{1/2}$. We therefore recorded the far-FTIR spectra of all the members of the [Fe($n$ditz)$_3$](ClO$_4$)$_2$ series; interestingly, almost no difference was observed (see Table 3). Strikingly, the LS state shows a band that only

Table 3. Temperature-dependent far-IR bands that can be associated with HS and LS species of [Fe($n$ditz)$_3$](ClO$_4$)$_2$.

n	HS	LS
5	469(w), 388(w), 367(m)	421(s), 380(w), 358(sh), 287(s)
6	476(m), 378(s), 357(sh), 282(sh)	422(m), 393(s), 364(m)
8	437(sh), 373(m), 338(m), 310(sh)	424(s), 395(w), 379(w), 354(s)
9	366(sh), 331(sh)	422(s), 354(m)

varies between 421 and 424 cm^{-1} and the HS state shows bands that vary by a maximum of 10 cm^{-1} and certainly do not follow any obvious trend. From that we can conclude that the shape of the potential wells of the LS and the HS states, as well as the horizontal distance between them (ΔQ), can be considered almost the same for all complexes. In other words, along the [Fe($n$ditz)$_3$](ClO$_4$)$_2$ series the change in $T_{1/2}$ implies a vertical displacement of the two potential wells and hence a variation in ΔE with the parity, as shown in Figure 7.

A further consequence of the variation of ΔE can be explained as it has previously been shown that there is a direct relationship between the rate constant of tunnelling, k_∞, and $T_{1/2}$.[26] Based on this finding we can see that if the stability of the photo-induced HS state decreases with the increase of $T_{1/2}$ (and of ΔE), for a constant intensity of light irradiation, the photo-excitation becomes more and more difficult, as observed experimentally (%Irr-Surface; reported in Table 2).

Photo-magnetic properties: The photo-magnetic properties of the different compounds were investigated by following the effect of irradiation with light under the influence of an applied magnetic field. At first, the sample was cooled slowly down to 10 K in order to stabilise the low-spin state; it was then irradiated and the change in magnetism was followed. When the saturation point was reached the light was switched off, the temperature was increased at a rate of 0.3 K min^{-1} and the magnetisation measured every 1 K. This procedure allows the quantification of T(LIESST), which is determined by the minimum of the $\partial\chi_\mathrm{M}T/\partial T$ versus T curve recorded during relaxation.[5] Figure 8 shows the photo-magnetic behaviour together with the magnetic susceptibility for each compound for the sake of completeness.

Table 2 collects, for each compound, the highest percentage of photo-conversion obtained by irradiation at 10 K relative to the magnetic value recorded at room temperature for a pure HS state (%Irr-Bulk). As expected from the surface analysis, we can see here that with bulk detection the level of photo-excitation is not uniform for all the compounds, even when special attention was paid to tune the intensity and/or the wavelength. This confirms once more that the level of photo-excitation decreases with an increase of $T_{1/2}$ and therefore clearly shows the influence of both the length and parity of the bridging ligand (see Figure 4).

If we now compare the shape of the T(LIESST) curve recorded by increasing the temperature, the complexes continue to behave differently from one other. For those with a short spacer, the $\chi_\mathrm{M}T$ product increases in the 10–30 K range, while for long spacers the magnetic signal strongly decreases. This behaviour provides new evidence that the stability of the photo-induced HS state in the tunnelling region varies along the [Fe($n$ditz)$_3$](ClO$_4$)$_2$ series. In fact, only for complexes possessing a sufficiently long-lived lifetime does the photo-induced HS fraction remain almost independent of any change of the temperature and of the effect of time during the measurement of the T(LIESST) curve. Consequently, the $\chi_\mathrm{M}T$ curve displays an increase in the magnetic response with the temperature due to the effect of the zero-field splitting (ZFS) of the iron(II) HS state in a non-perfectly octahedral geometry.[27] Indeed, the series of odd-numbered ligands show a ZFS effect that progressively vanishes in the order [Fe(5ditz)$_3$](ClO$_4$)$_2$ > [Fe(7ditz)$_3$](ClO$_4$)$_2$ > [Fe(9ditz)$_3$](ClO$_4$)$_2$, thereby reflecting a decrease of the lifetime of the photo-induced HS state in the tunnelling region.

An additional way to see the stability of the photo-induced HS state is to compare the magnitude of the T-(LIESST) temperatures.[5,6] Table 2 collects T(LIESST) temperatures of the [Fe($n$ditz)$_3$](ClO$_4$)$_2$ family, with the exception of the complexes [Fe(6ditz)$_3$](ClO$_4$)$_2$ and [Fe(8ditz)$_3$](ClO$_4$)$_2$, where the minimum on the $\partial\chi_\mathrm{M}T/\partial T$ versus T curve cannot be properly determined due to the very low efficiency of the photo-excitation (Figure 8). Nevertheless, it is interesting to see that the lowest T(LIESST) is found for [Fe(9ditz)$_3$](ClO$_4$)$_2$, which presents the highest $T_{1/2}$ value, as expected from the previous discussion of the stability of the photo-induced HS state.

Finally, it is important to note that the peculiarity of [Fe(4ditz)$_3$](ClO$_4$)$_2$, which is characterised by a two-step thermal spin-transition, shows similar features in the T(LIESST) experiments. Analysis of the $\partial\chi_\mathrm{M}T/\partial T$ versus T curve clearly shows the existence of two minima, at 39 K and 58 K, thus proving that both magnetically non-equivalent iron(II) metal centres can be photo-excited. It is also reasonable to propose that the T(LIESST) temperature found at 58 K corresponds to photo-excitation of the iron(II) metal centres involved in the LS/HS thermal spin-transition occurring at 84 K, and the T(LIESST) temperature at 39 K is linked to an SC phenomenon occurring at 134 K. The presence of two minima on the T(LIESST) curve is not unexpected as a similar result has been found for a mononuclear iron(II) SC material that also displays a two-step thermal spin-transition.[28]

Conclusion

We have presented the first systematic study of a homologous series of iron(II) spin-crossover coordination polymers

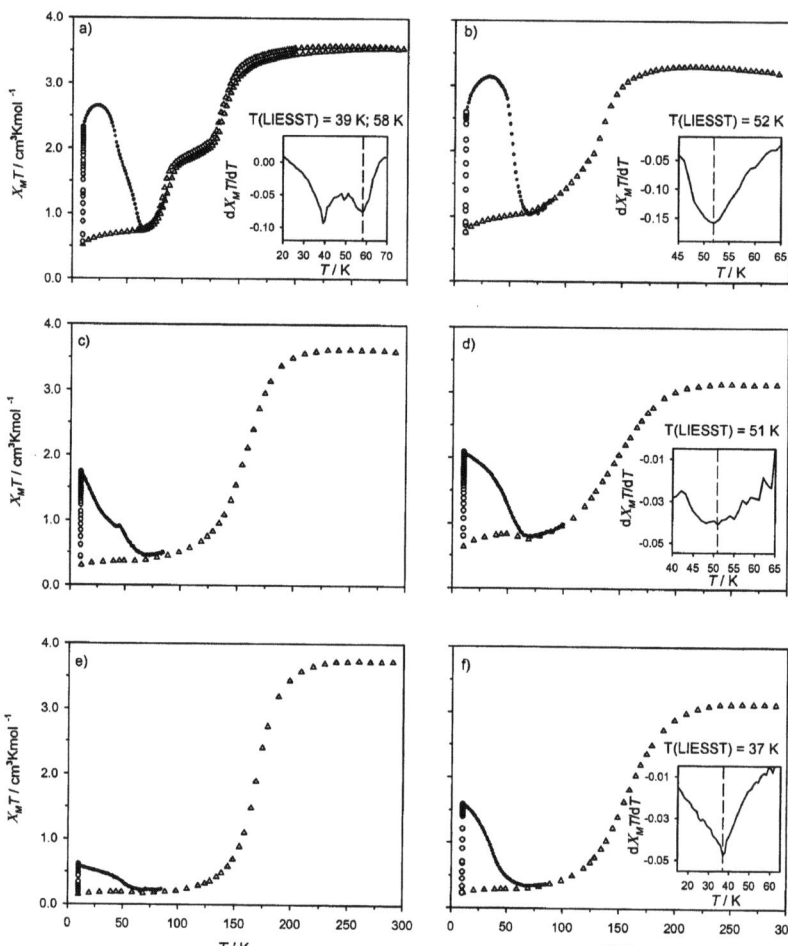

Figure 8. Temperature dependence of $\chi_M T$ for a) [Fe(4ditz)$_3$](ClO$_4$)$_2$, b) [Fe(5ditz)$_3$](ClO$_4$)$_2$, c) [Fe(6ditz)$_3$](ClO$_4$)$_2$, d) [Fe(7ditz)$_3$](ClO$_4$)$_2$, e) [Fe(8ditz)$_3$]-(ClO$_4$)$_2$ and f) [Fe(9ditz)$_3$](ClO$_4$)$_2$. ($\triangle$) Data recorded without irradiation; ($\circ$) data recorded with irradiation at 10 K; ($\bullet$) T(LIESST) measurement, data recorded in the warming mode with the laser turned off after irradiation for one hour. (Note that the small bump at 49 K observable in (c) is due to a small amount of dioxygen).

containing ClO$_4^-$ as counteranion and have demonstrated that for the [Fe($n$ditz)$_3$](ClO$_4$)$_2$ ($n=4$–9) series both the length and parity of the spacer have a large and systematic influence on the magnetic and photo-magnetic properties. With the exception of [Fe(4ditz)$_3$](ClO$_4$)$_2$, we have shown a two-step transition that is similar to the previously published compound obtained with PF$_6^-$ as anion.[14] This behaviour contrasts with [Fe(2ditz)$_3$(BF$_4$)$_2$], which is known to display an incomplete gradual thermal spin-transition at about 140 K.[18] A butylene group between the tetrazoles seems to provide the ideal distance between iron(II) centres that allows an interpenetrated network structure to be formed. In the case of longer spacers, the thermal spin-transition is also gradual, although whether this is due to structural changes is still unclear. Unfortunately, we have not been able to obtain single crystals of these coordination polymers.

From our magnetic studies, we have provided evidence to show that the thermal spin-transition temperature increases with the number of carbon atoms (n) in the spacer and we have reported a fascinating effect of the parity. The complexes with an even "n" display a more abrupt spin transition than odd-numbered ones. The $T_{1/2}$ values have also been found to be higher for the complexes having even-numbered $n$ditz ligands than for the odd-numbered ligands.

From our photo-magnetic investigations, we have also highlighted that the LIESST properties of the [Fe($n$ditz)$_3$]-(ClO$_4$)$_2$ family are strongly affected by the parity of the bridging ligands. Based on the SCC model in combination with far-FTIR spectroscopic data, we have proposed that the energy difference between the LS and the HS states is higher for the even series. Moreover, we have experimentally observed that the level of photo-excitation, as well as the T(LIESST) temperature, perfectly follow the tendency defined by $T_{1/2}$.

Acknowledgements

Financial support from the Austrian Science Foundation FWF (project 15874-N03) and the Italian MIUR, FIRB and PRIN projects is gratefully acknowledged. Furthermore, we thank the EU COST D14 action project 0011/01 for granting A. Absmeier a "Short Term Scientific Mission" in Bordeaux. Additionally, M. Bartel carried out research in Florence within the Marie Curie Training site LAMM (MOLMAG-MEST-CT-2004-504204).

[1] *Molecular Magnets: Recent Highlights* (Eds.: W. Linert, M. Verdaguer), Springer, Wien, **2003**.
[2] *Spin Crossover in Transition Metal Compounds Vols. I–III*, in *Topics in Current Chemistry* (Eds.: P. Gütlich and H. A. Goodwin), Springer, Berlin, **2004**.
[3] S. Decurtins, P. Gütlich, K. M. Hasselbach, H. Spiering, A. Hauser, *Inorg. Chem.* **1985**, *24*, 2174–2178.
[4] A. Hauser, P. Gütlich, H. Spiering, *Inorg. Chem.* **1986**, *25*, 4245–4248.
[5] J-F. Létard, L. Capes, G. Chastanet, N. Moliner, S. Létard, J.-A. Real, O. Kahn, *Chem. Phys. Lett.* **1999**, *313*, 115–120.
[6] J-F. Létard, P. Guionneau, O. Nguyen, J. S. Costa, S. Marcén, G. Chastanet, M. Marchivie, L. Goux-Capes, *Chem. Eur. J.* **2005**, *11*, 4582–4589.
[7] P. Poganiuch, S. Decurtins, P. Gütlich, *J. Am. Chem. Soc.* **1990**, *112*, 3270–3278.
[8] Th. Buchen, P. Gütlich, *Chem. Phys. Lett.* **1994**, *220*, 262–266.
[9] A. F. Stassen, O. Roubeau, I. F. Gramage, J. Linares, F. Varret, I. Mutikainen, U. Turpeinen, J. G. Haasnoot, J. Reedijk, *Polyhedron* **2001**, *20*, 1699–1707.
[10] G. Chastanet, A. B. Gaspar, J. A. Real, J. F. Létard, *Chem. Commun.* **2001**, 819–820.
[11] A. B. Gaspar, V. Ksenofontov, H. Spiering, S. Reiman, J. A. Real, P. Gütlich, *Hyperfine Interact.* **2002**, *144/145*, 297–306.
[12] V. Niel, A. Galet, A. B. Gaspar, M. C. Munoz, J. A. Real, *Chem. Commun.* **2003**, 1248–1249.
[13] J. G. Haasnoot, *Coord. Chem. Rev.* **2000**, *200–202*, 131–185.
[14] C. M. Grunert, J. Schweifer, P. Weinberger, W. Linert, K. Mereiter, G. Hilscher, M. Muller, G. Wiesinger, P. J. van Koningsbruggen, *Inorg. Chem.* **2004**, *43*, 155–165.
[15] Y. Sunatsuki, H. Ohata, M. Kojima, Y. Ikuta, Y. Goto, N. Matsumoto, S. Iijima, H. Akashi, S. Kaizaki, F. Dahan, J.-P Tuchagues, *Inorg. Chem.* **2004**, *43*, 4154–4171.
[16] C. M. Grunert, P. Weinberger, J. Schweifer, C. Hampel, A. F. Stassen, K. Mereiter, W. Linert, *J. Mol. Struct.* **2005**, *733*, 41–52.
[17] P. J. van Koningsbruggen, Y. Garcia, O. Kahn, L. Fournes, H. Kooijman, A. L. Spek, J. G. Haasnoot, J. Moscovici, K. Provost, A. Michalowicz, F. Renz, P. Gütlich, *Inorg. Chem.* **2000**, *39*, 1891–1900.
[18] J. Schweifer, P. Weinberger, K. Mereiter, M. Boca, C. Reichl, G. Wiesinger, G. Hilscher, P. J. van Koningsbruggen, H. Kooijman, M. Grunert, W. Linert, *Inorg. Chim. Acta* **2002**, *339*, 297–306.
[19] P. J. van Koningsbruggen, Y. Garcia, H. Kooijman, A. L. Spek, J. G. Haasnoot, O. Kahn, J. Linares, E. Codjovi, F. Varret, *J. Chem. Soc. Dalton Trans.* **2001**, *1*, 466–471.
[20] P. L. Franke, G. Haasnoot, A. P. Zuur, *Inorg. Chim. Acta* **1982**, *59*, 5–9.
[21] Bruker (2001). Programs SMART, version 5.054; SAINT, version 6.2.9; SADABS version 2.03; XPREP, version 5.1; SHELXTL, version 5.1. Bruker AXS Inc., Madison, Wisconsin, USA.
[22] G. M. Sheldrick, Programs SHELXS97 and SHELXL97, University of Göttingen, Germany, **1997**.
[23] C. Carbonera, A. Dei, C. Sangregorio, J.-F. Létard, *Chem. Phys. Lett.* **2004**, *396*, 198–201.
[24] M. Hostettler, K. W. Törnroos, D. Chernyshov, B. Vangdal, H.-B. Bürgi, *Angew. Chem.* **2004**, *116*, 4689–4695; *Angew. Chem. Int. Ed.* **2004**, *43*, 4589–4594.
[25] A. Hauser, *Coord. Chem. Rev.* **1991**, *111*, 275–290.
[26] A. Hauser in *Spin Crossover in Transition Metal Compounds III* (Eds.: P. Gütlich, H. A. Godwin), Springer, Berlin, **2004**, pp. 155–198.
[27] J.-F. Létard, G. Chastanet, O. Nguyen, S. Marcén, M. Marchivie, P. Guionneau, D. Chasseau, P. Gütlich, *Monatsh. Chem.* **2003**, *134*, 165–183.
[28] G. S. Matouzenko, J.-F. Létard, S. Lecocq, A. Boussezou, L. Capes, L. Salmon, M. Perrin, O. Kahn, A. Collet, *Eur. J. Inorg. Chem.* **2001**, 2935–2945.

Received: August 3, 2005
Published online: December 9, 2005

Field induced low-spin high-spin transition

B. Richter[a], A. Kirste[a], S. Hansel[a], M. von Ortenberg[a*], A. Absmeier[b], W. Linert[b], R. Groessinger[b]

[a]*Humboldt University at Berlin, Magnetotransport in Solids, Newtonstrasse 15, D-12489 Berlin, Germany*
[b]*Technical University Vienna, Karslplatz 13, A-1040 Vienna, Austria*

Elsevier use only: Received date here; revised date here; accepted date here

Abstract

We have investigated the temperature and magnetic field induced high-spin low-spin transition of [μ-Tris(1,4-bis(tetrazol-1-yl)butane-N4,N4´)iron(II)] Bis(hexafluorophosphate) and [μ-Tris(1,8-bis(tetrazol-1-yl)octane-N4,N4´)iron(II)] Bis(perchlorate) in transient megagauss fields using optical reflection measurements at λ=0.632 μm and λ=0.541 μm radiation. The experimental set-up uses POF in an environment between 140 K and 200 K. For sample temperature within this transition range magnetic field induced small reflection changes with strong hysteresis were detected. Evidently the strong (160 T) but short (5 μsec) magnetic field pulse cannot drive the transition into saturation. Using magnetic field pulses with different time constants the dynamic process could be studied.
For [μ-Tris(1,4-bis(tetrazol-1-yl)butane-N4,N4´)iron(II)] Bis(hexafluorophosphate) it was found that the solvent used in the actual production process of the material had a strong influence on the experimental results.
The project was supported by EuroMagNET.

© 2006 Elsevier B.V. All rights reserved

75.5; 75.30C;

low-spin high-spin transition; megagauss magnetization;

1. Introduction

Spin crossover in magnetic materials has become an increasingly investigated phenomenon in magnetism. Due to the different degeneracy of the high-spin and low-spin levels the majority population, and hence the characteristic features of the material, are changed as a function of temperature. This manifests not only in the magnetic properties but also in secondary features as optical properties [1]. Generally this crossover can also be induced by the different magnetic-field dependence of the spin levels involved, so that beyond a critical magnetic field the population of the spin-levels is inverted. In this way temperature, and magnetic field are complementary parameters to control the system with respect to the majority population and its related effects experimentally.

2. Materials

We have investigated the following two materials in powder form [μ-Tris(1,4-bis(tetrazol-1-yl)butane-N4,N4´)iron(II)] Bis(hexafluorophosphate) [2,3] and [μ-Tris(1,8-bis(tetrazol-1-yl)octane-N4,N4´)iron(II)] Bis(perchlorate) [3] . Both materials exhibit a pronounced visible color change in the temperature range between 77 K and 300 K indicating the high-spin/low-spin phase transition. The principal objective of the present experiments was the demonstration that this phase transition and hence change in color can also be obtained by application of strong magnetic fields in the megagauss range, since lower magnetic fields up to 40 T proved to be not successful for these two materials.

3. Experimental Setup

To detect the high-spin/low-spin transition we have applied a setup measuring the normal reflection in Faraday configuration using Plastic Optical Fiber (POF) for the monochromatic laser radiation of λ=632 nm or λ=541 nm. The corresponding sample holder was mounted in a miniature N_2-cryostat to meet the limited dimensions of the magnetic field coil. As detector we used a fast photo-diode with 125 MHz bandwidth.
As magnetic field generator we used the single-turn coil providing in our experiments peak fields of 160 T for a half-sine pulse of the order of 6 μsec length. The single-turn coils of 12 mm and 15 mm diameter were driven by a 225 kJ/60 kV

capacitor bank providing a peak current of the order 3 MA [4]. The magnetic field was measured by a calibrated pick-up coil with suitable integrator. Both data channels were set at 100 MHz sampling frequency.

4. Experimental Results

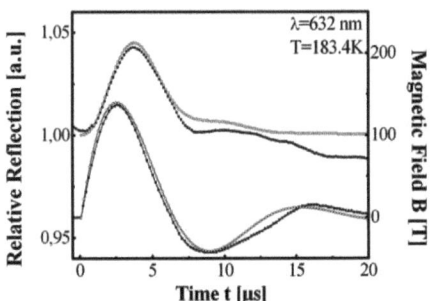

Fig. 1 Experimental data (solid curves) and simulation for the relative reflection and the magnetic field as a function of time.

Each of the mounted samples was checked for its exact temperature dependence of the phase transition to find the optimal temperature for the experiment in megagauss fields. In Fig. 1 we have plotted both the reflection data and the magnetic field intensity as a function of time in the upper and lower part respectively for [μ-Tris(1,4-bis(tetrazol-1-yl)butane-N4,N4')iron(II)] Bis(hexafluorophosphate) for λ=632 nm at T = 183.4 K. The second wavelength λ=541 nm provided no additional information.

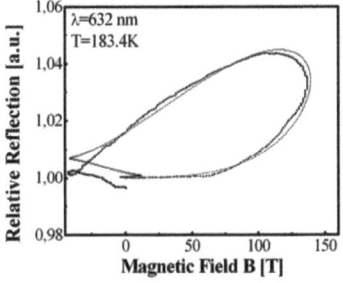

Fig. 2 Experimental data (solid) and simulation of the hysteris curve.

The experimental data of the reflection (upper solid curve in Fig.1) follows the experimental magnetic field (lower solid curve) with a pronounced hysteresis. This means, that the relaxation times involved are of the order of μsec. There is a clear increase of the reflectivity with increasing magnetic field indicating the increased population of the high-spin levels.

The broken curves indicate the results of the corresponding simulation discussed below. To demonstrate both the magnetic field dependence and the hysteresis involved we have plotted the data directly as a function of the applied magnetic field (Fig. 2). Again solid and broken curves indicate experimental and simulation results, respectively.

5. Discussion and Simulation

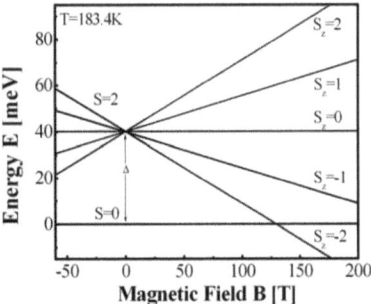

Fig. 3 Energy-level system as a function of the magnetic field.

To simulate the experimental results we have applied the model of a split S=2/S=0 level system as shown in Fig. 3 corresponding to the Hamiltonian [5]:

$$H = (\Delta + g\mu_B S_z B) \cdot \delta_{2,S} + 0 \cdot \delta_{0,S}$$

The dynamics of the system is determined by:

$$\frac{df(E_i)}{dt} = -\frac{(f(E_i) - f_0(E_i))}{\tau}$$

Here $f(E_i)$ and $f_0(E_i)$ are the non-equilibrium and equilibrium occupation probabilities of the levels E_i as indicated in Fig. 3. The parameters Δ, g, and τ were fit to the experimental data: Δ = 40 meV, g = 2.7 , τ = 1.6 μsec.
The value of Δ is within the expected range. Evidently the g-factor is increased by ligand interaction. The novel feature is the determination of the dynamical relaxation time τ.

6. Acknowledgment

The project was supported by EuroMagNET.

References

[1] P. Gütlich et al. *Angew.Chem.* **106** (1994), p. 2109
[2] C. M. Grunert et al. *Inorg. Chem.* **43** (2004), p. 155.
[3] A. Absmeier, M. Bartel, C. Carbonera, G.N.L. Jamenson, P. Weinberger, A. Caneschi, K. Mereiter, J.-F. Létard, W. Linert *Chem. Eur. J.* **12** (2006), p. 2235.
[4] O. Portugall et al., *J. Phys. D : Appl. Phys.* **32** (1999) p. 2354
[5] P. Schuster. „Ligandenfeldtheorie" *Verlag Chemie*, Weinheim, (1973).

Bibliography

[1] L. Cambi and L. Szego, *Chem. Ber. Dtsch. Ges.*, **1931**, 64, 2591.
L. Cambi and L. Szego, *Chem. Ber. Dtsch. Ges.*, **1933**, 66, 656.
L. Cambi and L. Malatesta, *Chem. Ber. Dtsch. Ges.*, **1937**, 70, 2067.

[2] C.D. Coryell, F. Stitt and L. Pauling, *J. Am. Chem. Soc.*, **1937**, 59, 633.

[3] L.Orgel, *Proc. 10th Solvay Conference*, ed. R. Stoops, Brussels, **1956**.

[4] P. George, J. Meetlestone and J.S. Griffith, *Haematin Enzymes*, Pergamon Press, New York, **1961**.

[5] (a) A.H. Ewald, R. L. Martin, I. Ross and A.H. White, *Proc. R. Soc. London, Ser. A*, **1964**, 280, 235; (b) A.H. White, R. Roper, E. Kokot, H. Waterman and R.L. Martin, *Austr. J. Chem.*, **1964**, 17, 294; (c) R.M. Golding, W.C. Tennant, C.R. Kanekar, R.l. Martin and A.H. White, *J. Chem. Phys*, **1966**, 45, 2688; (d) A.H. Ewald, R.L. Martin, E. Sinn, A.H. White, *Inorg. Chem.*, **1969**, 8, 1837.

[6] W.A. Baker and H.M. Bobonich,Jr., *Inorg. Chem.*, **1967**, 6, 48.

[7] E. König and K. Madeja, *Inorg. Chem.*, **1967**, 6, 48.

[8] a) K.A. Reeder, E.V. Dose, L.J. Wilson, *Inorg. Chem.*, **1978**, 17, 1071; b) M.S. Haddad, W.D. Federer, M.W. Lynch, D.N. Hendrickson, *Inorg. Chem.*, **1981**, 20, 131; c) H. Oshio, K. Kitazaki, J. Mishiro, N. Kato, Y.Maeda, Y. Takashima, *J. Chem. (Dalton Trans)*, **1987**, 1341; d) S. Schenker, A.R.M. Dyson, *Inorg.Chem.*, **1996**, 35, 4676.

[9] a) D.M. Halepoto, D.G.L. Holt, L.F. Larkworthy, G.J. Leigh, D.C. Povey, W. Smith, *J. Chem. Soc. Chem. Comm.*, **1989**, 18, 1322; b) M. Sorai, Y. Yumoto, D.M. Halepoto, L.F. Larkworthy, *J. Phys. Chem. Solids*, **1993**, 54(4), 421.

[10] a) J.H. Ammeter, R. Bucher, N. Oswald, *J. Am. Chem Soc.*, **1974**, 96, 7883; b) M.E. Switzer, R. Wang, M.F.Rettig, A.H. Maki, *J. Am. Chem. Soc.*, **1974**, 96, 7669; c) D. Cozak, F. Gauvin, *Organometallics*, **1987**, 6, 1912.

[11] R.C. Stoufer, D.H. Bush, W.B. Hardley, *J. Am. Chem. Soc.*, **1961**, 83, 3732.

[12] a) R.C. Stouter, D.W. Smith, E.A. Cleavenger, T.E. Norris, *Inorg. Chem.*, **1966**, 5, 1167; b) J. Zarembowitch, O.Kahn, *Inorg. Chem.*, **1984**, 23, 589; c) J. Zarembowitch, *New. J. Chem.*, **1992**,16, 255; d) J. Faus, M. Julve, F.Lloret, J.A. Real, J. Sletten, *Inorg. Chem.*, **1994**, 33, 5535; e) K. Heinze, G. Huttner, L. Zsolnai, P. Schober,*Inorg. Chem.* **1997**, 36, 5457.

[13] a) P.G. Sim, E. Sinn, *J. Am. Chem. Soc.*, **1981**, 103, 241; b) L. Kaustov, M.E. Tal, A.I. Shames, Z. Gross, *Inorg. Chem.*, **1997**, 36, 3503.

[14] a) W. Kläui, *J. Chem. Soc. Chem. Comm.*, **1979**, 700; b) P. Gütlich, B.R. Mc Garvey, W. Kläui, *Inorg. Chem.*,**1980**, 19, 3704; c) W. Eberspach, N. El Murr, W. Kläui, *Angew. Chem. Int. Ed. Engl.*, **1982**, 21, 915; d) G.Navon, W. Kläui, *Inorg. Chem.*, **1984**, 23, 2722; e) W. Kläui, W. Eberspach, P. Gütlich, *Inorg. Chem.*, **1987**, 26,3977.

[15] A. Hauser, *J. Chem. Phys.*, **1991**, 94, 2741.

[16] C. Joachim, J.K. Gimzewski and A. Aviram, *Nature*, **1974**, 408, 541.

[17] J.S. Griffith, *Proc. Roy. Soc.*, **1959**, 23, 23.

[18] C.J. Ballhausen, A.D. Liehr, *J. Am. Chem. Soc.*, **1959**, 81, 538.

[19] a) W.A. Baker, H.M. Bobonich, *Inorg. Chem.*, **1964**, 3, 1184; b) H.A. Goodwin, *Coord. Chem. Rev.*, **1976**, 18, 293.

Bibliography

[20] a) P. Gütlich, *Structure and Bonding* (Berlin), **1981**, 44, 83; b) L.F. Lindoy, S.E. Livingstone, *Coord. Chem.Rev.*, **1967**, 2, 173.

[21] a) M. Sorai, S. Seki, *J. Phys. Soc. Japan.*, **1972**, 33, 575; b) M. Sorai, S. Seki, *J. Phys. Chem. Solids*, **1974**, 35,555.

[22] M. Sorai, *J. Chem. Thermodynamics*, **2002**, 34, 1207-1253.

[23] E. König, G. Ritter, *Sol. State Comm.*, **1976**, 18, 279.

[24] M. Sorai, J. Ensling, K.M. Hasselbach, P. Gütlich, *Chem. Phys.*, **1977**, 20, 197.

[25] a) P.J. van Koningsbruggen, Y. Garcia, E. Codjovi, R. Lapouyade, O. Kahn, L. Fournès, L. Rabardel, *J. Mater.Chem.*, **1997**, 7, 2069 b) Y. Garcia, P.J. van Koningsbruggen, E. Codjovi, R. Lapouyade, O. Kahn, L. Rabardel,*J. Mater. Chem.*, **1997**, 7, 857.

[26] a) K.H. Sugiyarto, H.A. Goodwin, *Aust. J. Chem.*, **1988**, 41, 1645; b) K.H. Sugiyarto, D.C. Craig,A.D. Rae,H.A. Goodwin, *Aust. J. Chem.*, **1994**, 47, 869; c) K.H. Sugiyarto, K. Weitzner, D.C. Craig, H.A. Goodwin,*Aust. J. Chem.*, **1997**, 50, 869; d) K.H. Sugiyarto, M.L. Scuddler, D.C. Craig, H.A. Goodwin, *Aust. J. Chem.*,**2000**, 53, 75.

[27] H. Koeppen, E.W. Mueller, C.P. Koehler, H. Spiering, E. Meissner, P. Guetlich, *Chem. Phys. Lett.*, **1982**, 91, 348.

[28] *Spin Crossover in Transition Metal Compounds Vol I-III in Topics in Current Chemistry* (Eds. P. Gütlich and H. A. Goodwin), Springer, Berlin, **2004**.

[29] P. Gütlich, V. Ksenofontov, A.B. Gaspar, *Coord. Chem. Rev.*, **2005**, 249, 1811.

[30] A.Galet, A.B. Gaspar, G. Agusti, M. C. Munoz, G. Levchenko and J.A. real, *Eur. J. Inorg. Chem.*, **2006**, 3571.

[31] G. Molnar, T. Guillon, N.O. Moussa, L. rechignat, T. Kitazawa, M. Nardone, A. Bousseksou, *Chem. Phys. Let.*, **2006**, 423, 152.

[32] P. Gütlich, A. Hauser, H. Spiering, *Angew. Chem. Int. Ed. Engl.*, **1994**, 33, 2024.

[33] (a) V. Ksenofontov, A.B. Gaspar, G. Levchenko, B. Fitzsimmons, P.Gütlich, *J. Phys. Chem. B*, **2004**, 108, 7723; (b) P. Gütlich, A.B. Gaspar, V. Ksenofontov, Y. Garcia, *J. Phys. Cond. Matter.*, **2004**, 16, 1087.

[34] J. J. McGarvey and I. Lawthers, *J. Chem. Soc., Chem. Commun.*, **1982**, 906.

[35] S. Decurtins, P. Gütlich, C. P. Köhler, H. Spiering and A. Hauser, *Chem. Phys. Lett.*, **1984**, 105, 1.

[36] A. Hauser, *Chem. Phys. Lett.*, **1986**, 124, 543.

[37] P. Gütlich, Y. Garcia and H. A. Goodwin, *Chem. Soc. rev.*, **2000**, 29, 419.

[38] A. Hauser, *Coord. Chem. Rev.*, **1991**, 111, 275; A. Hauser, *Comments Inorg. Chem.*, **1995**, 17, 17.

[39] E. Buhks, G. Navon, M. Bixon and J. Jortner, *J. Am. Chem. Soc.*, **1980**, 102, 2918.

[40] a) J.-F.Létard, P. Guionneau, O. Nguyen, J. S. Costa, S. Marcén, G. Chastanet, M. Marchivie, L. Capes, *Chem. Eur. J.*, **2005**, 11, 4582. b) J.-F. Létard, *J. Mater. Chem.*, **2006**, 16, 2550.

[41] A. Desaix, O. Roubeau, J. Jeftic, J. G. Haasnoot, K. Boukheddaden, E.Codjovi, J. Linares, M. Nogues and F. Varret, *Eur. Phys. B.*, **1998**, 6, 183.

[42] F. Renz, H. Spiering, H. A. Goodwin and P. Gütlich, *Hyperfine Interact.*, **2000**, 126, 155.

[43] C. Roux, J. Zarembowitch, B. Gallois, T. Granier and R. Claude, *Inorg. Chem.*, **1994**, 33, 2273.

[44] N. Sasaki and T. Kambara, *J. Phys. Chem.*, **1982**, 15, 1035.

[45] Y. Qi, E. W. Müller, H. Spiering, and P. Gütlich, *Chem. Phys.Lett.*, **1983**, 101, 503.

[46] A. Bousseksou, N. Negre, M. Goiran, L. Salmon, J.-P. Tuchagues, M.-L. Boillot, and K. Boukheddaden, and F. Varret, *Eur. Phys. J.*, **2000**, B13, 451-456.

[47] J. G. Haasnoot, *Coord. Chem. Rev.* **2000**, 200-202, 131.

[48] S. Decurtins, P. Gütlich, K. M. Hasselbach, H. Spiering, A. Hauser, *Inorg. Chem.*, **1985**, 24, 2174.

[49] P. Poganiuch, S. Decurtins, P. Gütlich, *J. Am. Chem. Soc.*, **1990**, 112, 3270.

[50] Th. Buchen, P. Gütlich, *Chem. Phys. Lett.*, **1994**, 220 (3-5), 262.

[51] A. F. Stassen, O. Roubeau, I. F. Gramage, J. Linares, F. Varret, I. Mutikainen, U. Turpeinen, J. G. Haasnoot, J. Reedijk, *Polyhedron*, **2001**, 1699.

[52] C. M. Grunert, J. Schweifer, P. Weinberger, W. Linert, K. Mereiter, G. Hilscher, M. Muller, G. Wiesinger, P. J. van Koningsbruggen, *Inorg. Chem.*, **2004**, 43, 155.

[53] Y. Sunatsuki, H. Ohata, M. Kojima, Y. Ikuta, Y. Goto, N. Matsumoto, S. Iijima, H. Akashi, S. Kaizaki, F. Dahan, J-P Tuchagues, *Inorg. Chem.*, **2004**, 43, 4154.

[54] C. M. Grunert, P. Weinberger, J. Schweifer, C. Hampel, A. F. Stassen, K. Mereiter, W. Linert, *J. Mol. Struct.*, **2005**, 733, 41.

[55] P. J. van Koningsbruggen, Y. Garcia, O. Kahn, L. Fournes, H. Kooijman, A. L. Spek, J. G. Haasnoot, J. Moscovici, K. Provost, A. Michalowicz, F. Renz, P. Gütlich, *Inorg. Chem.*, **2000**, 39, 1891.

[56] J. Schweifer, P. Weinberger, K. Mereiter, M. Boca, C. Reichl, G. Wiesinger, G. Hilscher, P. J. van Koningsbruggen, H. Kooijman, M. Grunert, W. Linert, *Inorg. Chim. Acta*, **2002**, 339, 297.

[57] R.D.Shannon, *Acta Cryst.*,**1976**, A32, 751.

[58] P. J. van Koningsbruggen, Y. Garcia, H. Kooijman, A. L. Spek, J. G. Haasnoot, O. Kahn, J. Linares, E. Codjovi, F. Varret, *J. Chem. Soc. Dalton Trans.*, **2001**, 466.

[59] P. L. Franke, J. G. Hasnoot, A. P. Zuur, *Inorg. Chim. Acta.*, **1982**, 59, 5.

[60] Bruker (2001). Programs SMART, version 5.054; SAINT, version 6.2.9; SADABS version 2.03; XPREP, version 5.1; SHELXTL, version 5.1. Bruker AXS Inc., Madison, Wisconsin, USA.

[61] G. M. Sheldrick, Programs SHELXS97 and SHELXL97. University of Göttingen, , Göttingen, (Germany) **1997**.

[62] M. Bartel, *PhD Thesis, TU Wien*, **2007**.

[63] A. Absmeier, *Master Thesis, TU Wien*, **2004**.

[64] M. Takata,E. Nishibori,K. Kato, Y. Kubota, Y. Kuroiwa Y., M. Sakata, *Advances in X-ray Analysis*, **2002**, 45, 377.

[65] C. Carbonera, A. Dei, C. Sangregorio, J.-F. Létard, *Chem. Phys. Lett.*, **2004**, 396, 198.

[66] R.A. Brandt, Universität Duisburg NORMOS program, 1999.

[67] O.Portugall, N.Puhlmann, H.U.Müller, M.Barczewski, I.Stolpe, M.v.Ortenberg, *J. Phys. D: Appl. Phys 32*, **1999**, 18,2354.

[68] F.Herlach, *Rep. Prog. Phys. 62*, **1999**, 859.

[69] F.Herlach, M.v.Ortenberg, *IEEE Transactions on Magnetics*,**1996**, 32, 2438.

[70] H.P.Furth, M.A.Levine, R.W.Waniek, *Rev. Sci. Instrum.*, **1957**,28, 949.

[71] D.W.Forster, C.J.Martin, 2.5 megagauss from a capacitor discharge, Les Champs magnétiques intenses, leur Production et leurs Applications (Proc. Int. Conf., Grenoble), CNRS Paris 166, **1967**, 361.

[72] Maxwell, Inc., USA, US Patent Spezifikation

[73] J. W. Visser, *J. Appl. Cryst.*, **1969**, 2, 89.

[74] A. Le Bail, H. Duroy, J.L. Fourquet, *Mat. Res. Bull.*, **1988**, 23, 447.

[75] A.C. Larson, R.B. Von Dreele, General Structure Analysis System (GSAS), Los Alamos National Laboratory Report LAUR 86-748 (2004).

[76] P.W., Stephens, *J. Appl. Cryst.*, **1999**, 32, 281.

[77] R.J. Hill, p. 61 in R.A. Young (Ed.), *The Rietveld Method*, Oxford University Press, Oxford, **1993**.

[78] A. Boultif, D. Louër, *J. Appl. Cryst.*, **1991**, 24, 987.

[79] W. Linert, Diploma Thesis "' Struktur und Bindung in Eisen(II) Komplexverbindungen"', **1979**.

[80] J.-F. Létard, L. Capes, G. Chastanet, N. Moliner, S. Létard, J.-A. Real, O. Kahn, *Chem. Phys. Lett.*, **1999**, 313(1,2), 115.

[81] J.-F. Létard , G. Chastanet, O. Nguyen, S. Marcén, M. Marchivie, P. Guionneau, D. Chasseau, P. Gütlich, *Monatsh. Chem.*, **2003**, 134, 165.

[82] M. Yamada, E. Fukumoto, M. Ooidemizu, N. Bréfuel, N. Matsumoto, S. Iijjima, M. Kojima, N. Re, F. Dahan, J.-P. Tuchagues, *Inorg. Chem.*, **2005**, 44, 6967.

[83] Y. Ikuta, M. Ooidemizu, Y. Yamahata, M. Yamada, S. Osa, N. Matsumoto, S. Iijima, Y. Sunatsuki, M. Kojima, F.Dahan, J.-P. Tuchagues, *Inorg. Chem.*, **2003**, 42, 7001.

Bibliography

[84] Y. Sunatsuki, Y. Ikuta, N. Matsumoto, H. Ohta, M. Kojima, S. Iijima, S. Hayami, Y. Maeda, S. Kaizaki, F. Dahan, J.-P. Tuchagues, *Angew. Chem. Int. Ed.*, **2003**, 42, 1614.

[85] R. Hinek, H. Spiering, D. Schollmeyer, P. Gütlich and A. Hauser, *Chem. Eur. J.*,11,**1996**,1427.

[86] F. A.Cotton and G. Wilkinson: *Basic inorganic chemistry*, Wiley, **1987**.

[87] M. J. Winter: Oxford Chemistry Primers - *d*-Block Chemistry(27), **1994**.

[88] S.Sugano, Y. Tanabe and H. Kamimura, *Multiplets of Transition Metal Ions, Pure and Applied Physics 33*,Academic N.Y.,**1970**.

[89] H.L. Schäfer and G. Gliemann, *Einführung in die Ligandenfeldtheorie*, Akademische Verlagsgesellschaft, Wiesbaden, **1980**, 339.

[90] A. Hauser in *Spin Crossover in Transition Metal Compounds III*,(Eds.: P. Gütlich, H. A. Godwin), Springer, Berlin, **2004**, pp. 155.

[91] G. S. Matouzenko, J.-F. Létard, S. Lecocq, A. Boussezsou, L. Capes, L. Salmon, M. Perrin, O. Kahn, A. Collet, *Eur. J. Inorg. Chem.*, **2001**, 2935.

[92] J.-F. Létard, P. Guionneau, L. Rabardel, J. A. K. Howard,A.E. Goeta, D. Chasseau, O. Kahn, *Inorg. Chem.*, **1998**, 37, 4432.

[93] M. H. Klingele, B. Moubaraki, J. D. Cashion, K. S. Murray, S. Brooker, *Chem. Commun.*, **2005**, 8, 987.

[94] J.A. Real, I. Castro, A. Bousseksou, M. Verdaguer, R. Burriel, M. Castro, J. Linares, F. Varret, *Inorg. Chem.*, **1997**, 36, 455.

[95] V. Ksenofontov, H. Spiering, S. Reiman, Y. Garcia, A. B. Gaspar, N. Moliner, J. A. Real, P. Gütlich, *Chem. Phys. Lett.*, **2001**, 348, 381.

[96] P. Schuster *Ligandenfeldtheorie* Verlag Chemie, Weinheim,**1973**.

List of Figures

1.1. Presentation of the adiabatic potentials for the high-spin and the low-spin state along with the most important reaction coordinate for SC in Fe(II) complexes. 4

1.2. Principal types of spin-transition curves represented in the form of high-spin molar fraction, γ_{HS}, vs. temperature T: (a) gradual, (b) abrupt, (c) abrupt with hysteresis, (d) two-step and (e) incomplete. 6

1.3. $\chi_M T$ vs. T curves at different pressures for [Fe(*phen*)$_2$(NCS)$_2$] polymorph II [33]. 7

1.4. Schematic illustration of LIESST and reverse- LIESST of a d^6 complex in the SC range. Spin allowed d-d transitions are denoted by arrows and the radiationless relaxation processes by waved lines [37]. 9

1.5. [Fe(*phen*)$_2$(NCS)$_2$], the complete set of pulsed field experiments in the ascending (a) and descending (b) branches of the thermal hysteresis loop [46]. 11

1.6. The basic structural units of [Fe(4*ditz*)$_3$](PF$_6$)$_2$ showing how the ligands span the iron(II) centres to produce a three dimensional polymer. 13

2.1. Structural views of 5*ditz*, 7*ditz* and 9*ditz* (20% probability ellipsoids). All molecules have C_2 symmetry with the twofold axis passing through C4, C5 and C6, respectively. 4*ditz*, 6*ditz* and 8*ditz* are shown to the right for comparison [54] 21

List of Figures

2.2. Crystal structure of [Fe(4*ditz*)$_3$](BF$_4$)$_2$.EtOH at 89. Coordination environment of Fe(II). The disordered 4*ditz*-chain is drawn dashed, the unit cell is outlined. H atoms are omitted. 23

2.3. Crystal structure of [Fe(4*ditz*)$_3$](ClO$_4$)$_2$.EtOH at RT. Coordination environment of the two iron sites. H atoms are omitted. 24

2.4. Schematic view of the Reflectivity setup. With 1...helium dewar; 2...Control board; 3...lamp; 4...optical detector; 5...CVI spectrometer; 6...multimeter and 7...motorised sample holder   29

2.5. Scheme and characteristic points of a LIESST experiment   31

2.6. 12mm x 12mm x 3mm *scts* before the experiment (bottom), after a 10 kV, 6 kJ discharge with 37 T peak field (middle) and after a 55 kV, 189 kJ discharge with 188 T peak field (top).   32

2.7. Diagram of the Megagauss facility in Berlin   35

2.8. a) The induced magnetic fields depend on the dimension of the coil and the charging voltage b) Profile of the justified system with the sample holder, POF, Thermal element (TE) and the light conductor system.   36

3.1. Le Bail fitted X-ray powder patterns (λ ... Cu K$_{\alpha 1,2}$) of the 4*ditz* phases with BF$_4^-$, bottom and ClO$_4^-$, top. Calculated profiles are drawn with solid lines, reflection positions are marked with vertical ticks and the difference curves are shown at the bottom. Inset is the calculated diffractogram of [Fe(4*ditz*)$_3$](PF$_6$)$_2$.MeOH.   42

3.2. Comparison of the [Fe(n*ditz*)$_3$](BF$_4$)$_2$ (n = 5-10, 12, solid) series' X-ray powder patterns (λ ... Cu K$_{\alpha 1,2}$) with the corresponding perchlorates ($\cdots$), flanked by the corresponding SEM images (BF$_4^-$ left, ClO$_4^-$ right).   43

3.3. Synchrotron powder patterns of [Fe(5*ditz*)$_3$](BF$_4$)$_2$ at 300 K (–), 133 K (–), 50 K (–) and 10 K (–) (note that the measurement time was 60 min. in comparison to 5 min. in all other measurements).   44

List of Figures

3.4. Synchrotron powder patterns of [Fe(8*ditz*)$_3$](BF$_4$)$_2$ at 300 K (–), 250 K (–), 200 K (–), 150 K ( ), 100 K (–), 50 K (–), 9 K (–), 100 K (–) and 200 K (–). 45

3.5. Synchrotron powder patterns of [Fe(7*ditz*)$_3$](ClO$_4$)$_2$ at 300 K (–), 250 K (–), 200 K (–), 150 K ( ), 100 K (–), 50 K (–), 9 K (–), 100 K (–) and 200 K (–). 45

3.6. Temperature-dependent UV/VIS-NIR spectra of [Fe(9*ditz*)$_3$](ClO$_4$)$_2$ between 5000 and 35000 cm^{-1} and the corresponding *d-d* transitions. 46

3.7. Mole fraction of the optical measurement (squares) and magnetic measurement (dots) of [Fe(9*ditz*)$_3$](ClO$_4$)$_2$. The interception point corresponds to $T_{1/2}$. 48

3.8. Temperature dependent UV/VIS-NIR spectra of the BF$_4^-$ complexes with n = 4-9 (a - f) in the range of 5000 - 35000 cm^{-1}. Inserts represent the $^5T_2 \rightarrow\ ^5E$ transition. 51

3.9. Temperature dependent UV/VIS-NIR spectra of the ClO$_4^-$ complexes with n = 4-9 (a - f) in the range of 5000 - 35000 cm^{-1}. Inserts represent the $^5T_2 \rightarrow\ ^5E$ transition. 52

3.10. Presentation of the temperature dependent UV/VIS-NIR spectra in the range of 35000 - 5000 cm^{-1} of (a) [Fe(4*ditz*)$_3$](BF$_4$)$_2$, (b) [Fe(4*ditz*)$_3$](ClO$_4$)$_2$, (c) [Fe(4*ditz*)$_3$](PF$_6$)$_2$, (d) [Fe(4*ditz*)$_3$](SbF$_6$)$_2$ and (e) [Fe(4*ditz*)$_3$](ReO$_4$)$_2$. 55

3.11. Comparison of the [Fe(4*ditz*)$_3$](ClO$_4$)$_2$.MeOH "dry" (- -) and "wet" (–) complex measured at 101 and 120 K. 56

3.12. Reflectivity measurements of [Fe(5*ditz*)$_3$](ClO$_4$)$_2$ as a function of temperature. In the main figure the absorbance spectra at (–) 280 K; (- · ·) 140 K, (- · · -) 80 K, (· · ·) 60 K and (–) 10 K is shown. The insert reports the reflectivity followed at 550±2.5 nm and 830±2.5 nm. 58

List of Figures

3.13. Reflectivity of [Fe(4*ditz*)$_3$](BF$_4$)$_2$ followed at 830±2.5 nm as a function of temperature. Cooling (∇); warming ($\blacktriangle$). 60

3.14. Reflectivity measurement of [Fe(4*ditz*)$_3$](PF$_6$)$_2$. 61

3.15. Temperature dependency of the $\chi_M T$ product for the investigated complexes ($\blacktriangle$) [Fe(4*ditz*)$_3$](ClO$_4$)$_2$, ($\square$) [Fe(5*ditz*)$_3$](ClO$_4$)$_2$, ($\blacksquare$) [Fe(6*ditz*)$_3$](ClO$_4$)$_2$, ($\circ$) [Fe(7*ditz*)$_3$](ClO$_4$)$_2$, ($\bullet$) [Fe(8*ditz*)$_3$](ClO$_4$)$_2$ and ($\triangle$) [Fe(9*ditz*)$_3$](ClO$_4$)$_2$ b) cooling and heating modes for [Fe(4*ditz*)$_3$](ClO$_4$)$_2$. 64

3.16. Temperature dependency of the $\chi_M T$ product for the investigated complexes, a) ($\square$) [Fe(5*ditz*)$_3$](BF$_4$)$_2$, ($\blacksquare$) [Fe(6*ditz*)$_3$](BF$_4$)$_2$, ($\circ$) [Fe(7*ditz*)$_3$](BF$_4$)$_2$, ($\bullet$) [Fe(8*ditz*)$_3$](BF$_4$)$_2$ and ($\triangle$) [Fe(9*ditz*)$_3$](BF$_4$)$_2$ b) cooling and heating modes for [Fe(4*ditz*)$_3$](BF$_4$)$_2$. 67

3.17. Temperature dependence of $\chi_M T$ for (a) ($\blacktriangle$) [Fe(10*ditz*)$_3$](BF$_4$)$_2$ and ($\bullet$) [Fe(10*ditz*)$_3$](ClO$_4$)$_2$ b)($\triangle$) [Fe(12*ditz*)$_3$](BF$_4$)$_2$ and ($\circ$) [Fe(12*ditz*)$_3$](ClO$_4$)$_2$ in the heating mode. The existence of a small anomaly at around 50K on the curves almost certainly corresponds to an oxygen contamination. 67

3.18. Temperature dependence of $\chi_M T$ for (a) [Fe(4*ditz*)$_3$](ClO$_4$)$_2$, (b) [Fe(5*ditz*)$_3$](ClO$_4$)$_2$, (c) [Fe(6*ditz*)$_3$](ClO$_4$)$_2$, (d) [Fe(7*ditz*)$_3$](ClO$_4$)$_2$, (e) [Fe(8*ditz*)$_3$](ClO$_4$)$_2$, (f) [Fe(9*ditz*)$_3$](ClO$_4$)$_2$. ($\triangle$) Data recorded without irradiation; ($\circ$) data recorded with irradiation at 10 K; ($\bullet$) T *(LIESST)* measurement, data recorded in the warming mode with the laser turned off after irradiation for one hour. (Note that the small bump at 49 K observable in (c) is due to a small amount of oxygen). 70

List of Figures

3.19. Temperature dependence of $\chi_M T$ for (a) [Fe(4*ditz*)$_3$](BF$_4$)$_2$, (b) [Fe(5*ditz*)$_3$](BF$_4$)$_2$, (c) [Fe(6*ditz*)$_3$](BF$_4$)$_2$, (d) [Fe(7*ditz*)$_3$](BF$_4$)$_2$, (e) [Fe(8*ditz*)$_3$](BF$_4$)$_2$, (f) [Fe(9*ditz*)$_3$](BF$_4$)$_2$. ($\triangle$) Data recorded without irradiation; ($\circ$) data recorded with irradiation at 10 K; ($\bullet$) *T (LIESST)* measurement, data recorded in the warming mode with the laser turned off after irradiation for one hour. The existence of a small anomaly at around 50 K on T(LIESST) curve corresponds to an oxygen contamination even if particular precaution has been taken to purge the SQUID cavity for one hour at room temperature. 72

3.20. ^{57}Fe-Mössbauer spectra of [Fe(4*ditz*)$_3$](BF$_4$)$_2$ at selected temperatures. 74

3.21. Experimental data (– and –) and simulation (– and –) for the relative reflection and the magnetic field as a function of time. . 76

3.22. Experimental data (–) and simulation of the hysteresis. 76

3.23. Effect of the induced magnetic field on [Fe(4*ditz*)$_3$](PF$_6$)$_2$.EtOH complex at different temperatures. 77

3.24. Excitation of the SC in [Fe(4*ditz*)$_3$](PF$_6$)$_2$.EtOH at 183.4 K at different induced magnetic fields with the coil of 12mm× 12mm× 3mm and different charging voltages: (–) 30 kV,(–) 35 kV,(–) 40 kV,(–) 45kV and (–) 50 kV. 78

3.25. Excitation of the SC in [Fe(4*ditz*)$_3$](PF$_6$)$_2$.EtOH at 183.4 K at different induced magnetic fields with the coil of 13mm× 13mm× 3mm and different charging voltages: (–) 35 kV,(–) 40 kV,(–) 45 kV and (–) 50 kV. 79

3.26. Excitation of the SC in [Fe(4*ditz*)$_3$](PF$_6$)$_2$.EtOH at 183.4 K at different induced magnetic fields with the coil of 15mm× 15mm× 3mm and different charging voltages: (–) 40 kV,(–) 45 kV,(–) 50 kV and (–) 50 kV. 79

List of Figures

4.1. Tentative basic structural model of the [Fe(n*ditz*)$_3$](X)$_2$.EtOH (n = 5-10, 12, X = BF$_4^-$ and ClO$_4^-$) phases viewed along the [Fe(n*ditz*)$_3$]$^{2+}$ chains (circled, top) and the perpendicular direction (bottom). The orientation and geometry of the complex is presumed arbitrarily. 83

4.2. Absorption spectra of the [CoF$_6$]$^{-3}$ in K$_2$Na[CoF$_6$], the splitting of the t$_{2g}^3$e$_g^3$ excited state as a result of the Jahn-Teller distortion is visible [86] 85

4.3. Dependence between n and the width of the HS band for a) [Fe(n*ditz*)$_3$](ClO$_4$)$_2$ with n = 5-9 and b) [Fe(n*ditz*)$_3$](BF$_4$)$_2$ with n = 5-9 complexes. 86

4.4. $T_{1/2}$(a) and %Irr-Bulk (b) as a function of the number of carbons n in the spacer of the ditetrazole ligands for the perchlorate series. 90

4.5. Potential wells for an odd and even ditetrazole coordination polymer demonstrating the effect of the parity of the bridging ligand. Only ΔE appears to be affected and not ΔQ. 91

4.6. Comparison of $T_{1/2}$ as function of the number of carbons (n) in the spacer of the ditetrazole ligands for the BF$_4^-$ series. 94

4.7. Comparison of the mole fraction of [Fe(4*ditz*)$_3$](BF$_4$)$_2$ from SQUID (slow cooling ▽; slow warming △; warming preceded by rapid cooling ○) and Mössbauer measurement warming preceded by rapid cooling (●). 97

4.8. Energy-level system as a function of the magnetic field. 99

List of Figures

5.1. Temperature dependence of $\chi_M T$ for [Fe(n*ditz*)$_3$](BF$_4$)$_2$ with n = 5-10 and 12, [Fe(10*ditz*)$_3$](ClO$_4$)$_2$ and [Fe(12*ditz*)$_3$](ClO$_4$)$_2$ in the heating mode close to the transition temperatures. The insert shows the slope of the inflection point of the transition curves. BF$_4^-$ compounds are denoted with (•) [Fe(10*ditz*)$_3$](ClO$_4$)$_2$ and [Fe(12*ditz*)$_3$](ClO$_4$)$_2$ are denoted with (▲).The line connecting the points has no physical meaning but is intended as a guide to the eyes. 103

5.2. Comparison of $T_{1/2}$ as function of the number of carbons (n) in the spacer of the ditetrazole ligands for the BF$_4^-$ (▲) and ClO$_4^-$ (•) series. 104

A.1. Reflectivity measurements of the [Fe(n*ditz*)$_3$](BF$_4$)$_2$ complexes with n = 4-9. 108

A.2. Reflcetivity measurements of the [Fe(n*ditz*)$_3$](ClO$_4$)$_2$ complexes with n = 4,6-9 and 12. 109

A.3. Reflectivity measurements of the [Fe(n*ditz*)$_3$](PF$_6$)$_2$ complexes with n = 7-9. 110

A.4. Reflectivity measurements of the [Fe(n*ditz*)$_3$](SbF$_6$)$_2$ complexes with n = 7-9. 111

List of Tables

2.1. Yield, NMR and IR data of 5*ditz*, 7*ditz* and 9*ditz*. 19
2.2. Crystallographic data of 5*ditz*, 7*ditz* and 9*ditz*. 20
2.3. Yield, elemental analysis and mid-FTIR data of the [Fe(n*ditz*)$_3$](BF$_4$)$_2$ complexes. 25
2.4. Yield, elemental analysis and mid-FTIR data of the [Fe(n*ditz*)$_3$](ClO$_4$)$_2$ complexes. 26
2.5. Correlation between the coil dimensions and the induced magnetic fields. 33

3.1. Summery of the displacement of the HS band maximum (RT $\rightarrow$ LT) for the [Fe(n*ditz*)$_3$](BF$_4$)$_2$ and [Fe(n*ditz*)$_3$](ClO$_4$)$_2$ (n = 4-9) complexes. 50
3.2. Widths of the $^5T_2 \rightarrow {}^5E$ transition at RT and LT for the [Fe(n*ditz*)$_3$](BF$_4$)$_2$ and [Fe(n*ditz*)$_3$](ClO$_4$)$_2$ (n = 4-9) complexes. . 50
3.3. Δ_O values for LS and HS for the BF$_4^-$ and ClO$_4^-$ complexes. . . 53
3.4. Displacement of the HS band maximum between RT and LT, width of the $^5T_2 \rightarrow {}^5E$ transition at RT and LT and Δ_O LS respectively Δ_O HS for the [Fe(4*ditz*)$_3$](X)$_2$ (X= BF$_4^-$, ClO$_4^-$, PF$_6^-$, SbF$_6^-$ and ReO$_4^-$) complexes. 54
3.5. % Irradiation surface caculated from the reflectivity measurements of the discussed complexes 60
3.6. Temperature dependent far IR bands that can be associated with HS and LS species of [Fe(n*ditz*)$_3$](ClO$_4$)$_2$. 62
3.7. Spin transition and LIESST parameters of the discussed complexes. 65

List of Tables

3.8. Isomer shift (δ) and Quadrupole Splitting (Δ) obtained for [Fe(4*ditz*)$_3$](BF$_4$)$_2$. 74

Lebenslauf

Persönliche Daten

Name	Alina Absmeier
Adresse	Lenaustrasse 21, 4650 Lambach
Telefon	+43-676-4316403
email	absmeier@mail.zserv.tuwien.ac.at
Geburtsdatum	29. Juli 1979
Geburtsort	80801 München
Staatsangehörigkeit	Österreich / Deutschland
Familienstand	Ledig

Ausbildung

1990-1996	Mathem. -Naturwissenschaftliches Gymnasium München
1998	Matura im Juni 1998 am Realgymnasium der Benediktiner, 4650 Lambach
1998 - 11/2004	Studium der Technischen Chemie an der TU Wien. Abschluss mit dem akademischen Grad Dipl.-Ing.
Februar 2005	Short Term Scientific Mission an der Universität Bordeaux (Frankreich) in der Gruppe von Dr. Jean-Francois Létard im Rahmen des EU-COST Projektes D14/0011/01.
Oktober 2005	Aufenthalt an der Humboldt Universität zu Berlin im Rahmen des EuroMagNet Projektes.
November 2005	Übernahme der Handelsagentur der Mutter
01/2005-03/2007	Doktoratsstudium

Wien, am 19. Februar 2007

Die VDM Verlagsservicegesellschaft sucht für wissenschaftliche Verlage abgeschlossene und herausragende

Dissertationen, Habilitationen, Diplomarbeiten, Master Theses, Magisterarbeiten usw.

für die kostenlose Publikation als Fachbuch.

Sie verfügen über eine Arbeit, die hohen inhaltlichen und formalen Ansprüchen genügt, und haben Interesse an einer honorarvergüteten Publikation?

Dann senden Sie bitte erste Informationen über sich und Ihre Arbeit per Email an *info@vdm-vsg.de*.

Sie erhalten kurzfristig unser Feedback!

VDM Verlagsservicegesellschaft mbH
Dudweiler Landstr. 99   Telefon +49 681 3720 174
D - 66123 Saarbrücken   Fax   +49 681 3720 1749

www.vdm-vsg.de

Die VDM Verlagsservicegesellschaft mbH vertritt

Printed by Books on Demand GmbH, Norderstedt / Germany